Principles of Genome Analysis

A Guide to Mapping and Sequencing DNA from Different Organisms

S. B. PRIMROSE

b

Blackwell Science

© 1995 by
Blackwell Science Ltd
Editorial Offices:
Osney Mead, Oxford OX2 0EL
25 John Street, London WC1N 2BL
23 Ainslie Place, Edinburgh EH3 6AJ
238 Main Street, Cambridge
  Massachusetts 02142, USA
54 University Street, Carlton
  Victoria 3053, Australia

Other Editorial Offices:
Arnette Blackwell SA
  1, rue de Lille, 75007 Paris
  France

Blackwell Wissenschafts-Verlag GmbH
  Kurfürstendamm 57
  10707 Berlin, Germany

  Feldgasse 13, A-1238 Wien
  Austria

First published 1995

Set by Setrite Typesetters Ltd, Hong Kong
Printed and bound in Great Britain at
The Alden Press

DISTRIBUTORS

Marston Book Services Ltd
  PO Box 87
  Oxford OX2 0DT
  (Orders: Tel: 01865 791155
          Fax: 01865 791927
          Telex: 837515)

North America
  Blackwell Science, Inc.
  238 Main Street
  Cambridge, MA 02142
  (Orders: Tel: 800 215-1000
          617 876-7000
          Fax: 617 492-5263)

Australia
  Blackwell Science Pty Ltd
  54 University Street
  Carlton, Victoria 3053
  (*Orders*: Tel: 03 347-0300
          Fax: 03 349-3016)

A catalogue record for this title
is available from the British Library

ISBN 0-86542-946-4

Library of Congress
Cataloging-in-Publication Data

Primrose, S.B.
    Principles of genome analysis: a guide to
  mapping and sequencing DNA from
  different organisms/S.B. Primrose.
        p.   cm.
    Includes bibliographical references and
  index
    ISBN 0-86542-946-4
    1. Gene mapping.   2. Nucleotide
sequence.   I. Title.
    QH445.2.P75   1995
    574.87'322——dc20

Contents

Preface

Since the beginning of this century, a central problem in genetics has been the creation of maps of whole chromosomes. These maps are crucial for the understanding of the structure of genes, their function and their evolution. Until recently these maps were created by genetic means, i.e. as a result of sexual crosses. In the last 10 years the widespread use of recombinant DNA technology has permitted the generation of molecular or physical maps defined here as the ordering of distinguishable DNA fragments by their position along the chromosome. In some instances these physical maps can be as detailed as the complete DNA sequence of entire chromosomes.

Physical mapping of a wide range of genomes has occurred with incredible speed. Already a wide range of manipulative and analytical repertoires exist along with specialist journals and a specialist language. This means that it is very difficult for the experienced geneticist or molecular biologist entering the field to comprehend the latest developments or even what has been achieved. The aim of this text is to provide a general overview of the methodology and rationale employed. In such a fast-moving field it is difficult to be completely up-to-date but for those who worry about such a thing the literature available as late as March 1995 has been surveyed.

A number of individuals provided much appreciated assistance during the preparation of the manuscript and must be acknowledged here. In particular, Alan Hamilton, Hans Lehrach, Ian Dunham and John Armour read the entire text and made many helpful suggestions for improving it. Most of their recommendations were incorporated but any errors or omissions which remain are entirely my responsibility. Thanks also are due to Vera Butterworth and Lynne Goodman for assistance in finding and checking many of the references and to Margaret Courtney for typing the manuscript and then retyping it *ad nauseum*.

Abbreviations

AFLP	amplified fragment length polymorphism
APP	amyloid precursor protein
ARS	autonomously replicating sequence
BAC	bacterial artificial chromosome
bp	base pair
CAPS	cleaved amplified polymorphic sequences
CEPH	Centre d'Etude du Polymorphisme Humain
cM	centimorgan
ct	chloroplast
DIRVISH	direct visual hybridization
DMD	Duchenne muscular dystrophy
EMC	enzyme mismatch cleavage
ES	embryonic stem (cells)
EST	expressed sequence tag
FACS	fluorescence activated cell sorting
FISH	fluorescent *in situ* hybridization
GDRDA	genetically directed representational difference analysis
GMS	genome mismatch scanning
GSS	genome sequence sampling
HAEC	human artificial episomal chromosome
HPRT	hypoxanthine phosphoribosyl transferase
kb	kilobase
LINE	long interspersed nuclear element
LOD	logarithm_{10} of odds
LTR	long terminal repeat
MAC	mammalian artificial chromosome
MALDI/TOF/MS	matrix-assisted laser desorption ionization/time of flight/mass spectrometry
Mb	megabase
MHC	major histocompatibility complex
mt	mitochondrial
ORF	open reading frame
PAC	P1 artificial chromosome
PCR	polymerase chain reaction
PFGE	pulsed field gel electrophoresis

1 Rationale for mapping and sequencing genomes

Introduction

Currently a whole series of international efforts is underway to construct genetic and physical maps and to sequence the genomes of a diversity of organisms ranging from the bacterium *Escherichia coli* to humans. The task is a Herculean one. On a global basis it is occupying thousands of scientists and is costing hundreds of millions of dollars annually. Given the effort being expended, the task clearly is a complex one. So, how is the task being approached and why is it being undertaken in the first place? This chapter shall focus mainly on the second question, i.e. the rationale for mapping and sequencing genomes. The remainder of the book will concentrate on the methodology for mapping and sequencing genomes and the application of the knowledge gained.

Mapping genes and genomes

Mapping genes and the creation of genetic maps is a fundamental part of the science of genetics. This is because much of genetics is concerned with the understanding and manipulation of the inheritance of particular traits. For plants and animals of agronomic importance this means selective breeding and the identification of those offspring with the desired combination of characteristics. In the case of humans the objective is to predict whether the fetus carries genes for important inherited disorders, i.e. prenatal detection of genetic diseases. Where traits are associated with particular genes the task is relatively easy. Unfortunately there are many times more traits than identified genes and so geneticists make do with marker genes. These are genes which can easily be identified and which are genetically linked to the gene for the trait of interest.

In order to map the locus of a trait by genetic linkage, a panel of markers is tested in turn for evidence of co-segregation with the trait at meiosis. Two loci A and B are genetically linked if the alleles present at those loci on a particular chromosome tend to be transmitted together through meiosis. To be linked it is necessary, but not sufficient, for loci to be syntenic, i.e. on the same chromosome. The

1

combination of alleles at linked loci is called a haplotype; for example, haplotype A1B1 means a single chromosome carrying allele A1 at locus A and allele B1 at locus B. During meiosis, each pair of homologous chromosomes undergoes at least one recombination (cross-over) between non-sister chromatids. To show genetic linkage, loci must be located in close physical proximity on a chromosome. Thus, to map a new gene it is necessary to have a large number of different markers, ideally evenly spaced along each chromosome. In micro-organisms such as *E. coli* and yeast, such markers can be generated quite easily by classical mutation techniques. Moving up the evolutionary tree, the generation of markers becomes increasingly difficult and in man is ethically unacceptable. Instead, naturally occurring polymorphisms are used. By definition, all polymorphisms occur at the level of DNA but unfortunately few of them are readily scorable like the phenotypes used by Mendel in his classical work on peas. The advent of recombinant DNA technology suggested a completely new approach to defining potentially large numbers of marker loci: the use of DNA probes to identify polymorphic DNA sequences (Botstein *et al.* 1980). The first such DNA polymorphisms to be detected were differences in the length of DNA fragments after digestion with sequence specific restriction endonucleases, i.e. restriction fragment length polymorphisms (RFLPs; Fig. 1.1).

To generate an RFLP map the probes must be highly informative. This means that the locus must not only be polymorphic, it must be *very* polymorphic. If enough individuals are studied any randomly selected probe will eventually discover a polymorphism. However, a polymorphism in which one allele exists in 99.9% of the population and the other in 0.1% is of little utility since it seldom will be informative. Thus, as a general rule, the RFLPs used to construct the genetic map should have two, or perhaps three, alleles with equivalent frequencies.

The first RFLP map of an entire genome (Fig. 1.2) was that described for the human genome by Donis-Keller *et al.* (1987). They tested 1680 clones from a phage library of human genomic DNA to see whether they detected RFLPs by hybridization to Southern blots of DNA from five unrelated individuals. DNA from each individual was digested with 6–9 restriction enzymes. Over 500 probes were identified that detected variable banding patterns indicative of polymorphism. From this collection, a subset of 180 probes detecting the highest degree of polymorphism was selected for inheritance studies in 21 CEPH families (see below). Additional probes were generated from chromosome-specific libraries such that ultimately 393 RFLPs were selected. The various loci were arranged into linkage groups representing the 23 human chromosomes by a combination of mathematical linkage analysis and physical location of selected clones. The latter was achieved by hybridizing probes to panels of rodent – human hybrid cells containing varying human chromosomal

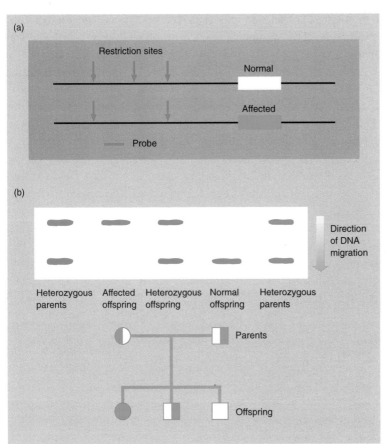

Figure 1.1 Example of a RFLP and its use for gene mapping. (a) A polymorphic restriction site is present in the DNA close to the gene of interest. In the example shown, the polymorphic site is present in normal individuals but absent in affected individuals. (b) Use of the probe shown in Southern blotting experiments with DNA from parents and progeny for the detection of affected offspring.

complements (see p. 45). RFLP maps have not been restricted to the human genome. For example, RFLP maps have been published for a number of plants, including maize (Coe *et al*. 1990; Burr & Burr 1991), *Arabidopsis* (Chang *et al*. 1988; Nam *et al*. 1989) and rice (Kurata *et al*. 1994).

The human genome map produced by Donis-Keller *et al*. (1987) was a landmark publication. However, it identified RFLP loci with an average spacing of 10 centimorgans (cM). That is, the loci had a 10% chance of recombining at meiosis. Given that the human genome is 4000 cM in length, the distance between the RFLPs is 10Mb on average. This is too great to be of use for gene isolation. However, if the methodology of Donis-Keller *et al*. (1987) was used to construct a 1 cM map, then 100 times the effort would be required! This is because ten times as many probes would be required and ten times more families studied. The solution has been to use more informative polymorphic markers and other mapping techniques and these are described in detail in Chapter 4. Use of these techniques has led to the generation of a human map in which 50% of the genome has markers at a distance apart of less than 1cM (Gyapay *et al*. 1994). More importantly, these advances in gene mapping have led to

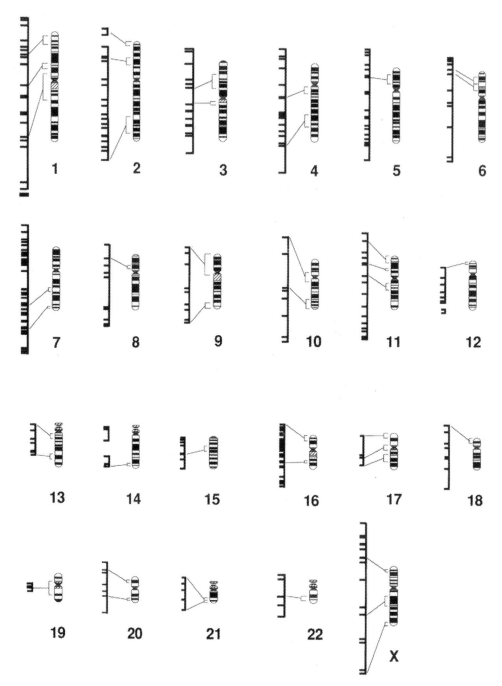

Figure 1.2 RFLP genetic linkage map of the human genome. (Reproduced with permission from Donis-Keller *et al.* 1987, copyright (1987) Cell Press.)

increased emphasis in developing representative genetic maps for a wide range of species, particularly domestic plants and animals, e.g. rice (Kurata *et al.* 1994) and pigs (Archibald 1994b).

It should be noted that man represents an extreme case of difficulty in creating a genetic map. Not only are directed matings not possible, but the length of the breeding cycle (minimum 15–16 years) makes

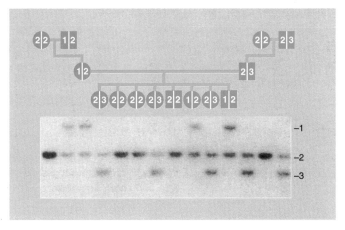

Figure 1.3 Inheritance of a RFLP in a CEPH family. The RFLP probe used detects a single locus on human chromosome 5. In the family shown, three alleles are detected on Southern blotting after digestion with *Taq*I. For each of the parents it can be inferred which allele was inherited from the grandmother and which from the grandfather. For each child the grandparental origin of the two alleles can then be inferred. (Redrawn with permission from Donis-Keller *et al.* 1987, copyright (1987) Cell Press.)

conventional analysis impractical. Consequently, the Centre d'Etude du Polymorphisme Humain (CEPH) was organized in Paris in 1984. The CEPH maintains cell lines from three-generation human families, consisting in most cases of four grandparents, two parents and an average of eight children (Dausset *et al.* 1990). Originally cell lines from 40 families were kept but the number now is much larger. Such families are ideal for genetic mapping because it is possible to infer which allele was inherited from which parent (Fig. 1.3). The CEPH distributes DNAs from these families to collaborating investigators around the world.

Understanding the phenotype

Using classical genetics it is relatively easy to show the mode of inheritance of a particular Mendelian trait, i.e. dominant vs. recessive, autosomal vs. sex linked. A full understanding of the biological basis of the phenotype requires a detailed knowledge of the appropriate gene(s), the genetic control of the gene(s) and identification of the gene product and its role in the life of the host. In some cases the jump from phenotype to gene analysis and biochemical explanation is relatively simple. For example, amino-acid auxotrophy and antibiotic resistance in micro-organisms or phenylketonuria and glucose-6-phosphate dehydrogenase (G6PD) deficiency in man are phenotypes which are easy to analyse biochemically. However, for the vast majority of traits, including over 4000 in man, no biochemical information is available. Recent developments in mapping and cloning have provided a solution, originally known as *reverse genetics* but now called *positional cloning*. In essense, one maps the gene, clones it, sequences it and then deduces function. The methodology for doing this is described in Chapter 6. Among the successes achieved with the use of positional cloning, the most significant must be the determination of the biochemical deficiency responsible for cystic fibrosis. The genetics of the disease were well understood (autosomal recessive) and the phenotype (viscid secretions, chest infections, limited lifespan) easily recognizable. However, until the gene was mapped, cloning and sequencing were not possible. Once sequence analysis could be undertaken it was relatively easy to show that cystic fibrosis is due to deletion of a phenylalanine codon in a gene encoding a chloride transport protein.

Comparative genome mapping

As noted above, recent advances in gene mapping have led to the development of representative genetic maps for a wide range of species. Initially these maps were compiled to provide a resource for genetic analysis in the species selected. More recently it has been realized that comparison of maps from related species has two benefits. First, by identifying conserved loci in related species and using them as reference points it is possible to transfer linkage information from 'map-rich' species (e.g. humans and mice) to 'map-poor' species (e.g. cows, pigs and sheep). Among mammals, most of the differences in karyotype are related to fixation of less than 100 reciprocal translocations (Edwards 1994). In the plant world comprehensive genetic maps are now available for all the world's important crop species. They show a remarkable conservation of order of markers over family-wide taxonomic groupings (e.g. rice, wheat and other cereals) (Schwarzacher 1994). Second, comparison of the maps should help dissect the evolution of genome organization and provides clues about the adaptive rationale, if any, behind particular structural arrangements.

For the purposes of comparative mapping, highly polymorphic loci are not particularly useful because they are seldom conserved sufficiently to recognize locus homology between different biological genera or orders. Coding gene loci tend to be more highly conserved between distantly related species and so are useful for comparing linkage. O'Brien *et al.* (1993) have selected a group of 321 coding gene loci suitable for comparative gene mapping in mammals and other vertebrates. All the loci selected are human and mouse functional genes which have been cloned and they are spaced on average 5–10 cM throughout their respective genomes.

A different use of comparative gene mapping is the development of mouse models of human genetic diseases. Some examples are given in Table 1.1 and a fuller listing has been provided by Copeland *et al.* (1993). The similarity of the mouse and human genetic maps offers two benefits. First, there are many mouse mutants that have been identified and which are thought to represent models for human diseases but which have not been cloned. The development of a high-density mouse linkage map should facilitate their cloning. If the human counterpart already has been cloned then it will facilitate the development of a mouse disease model. Second, mapping genes in the mouse can facilitate the cloning of human genetic disease loci,

Genetic alteration	Human disease equivalent
Introduction of mutant collagen gene into wild-type mice	Osteogenesis
Inactivation of mouse gene coding hypoxanthine-guanine phosphoribosyl transferase (HPRT)	HPRT deficiency
Mutation of locus for X-linked muscular dystrophy	X-linked muscular dystrophy
Introduction of activated human *ras* and *c-myc* oncogenes	Induction of malignancy
Introduction of mutant (Z) allele of human α_1-antitrypsin gene	Neonatal hepatitis
Introduction of HIV *tat* gene	Kaposi's sarcoma
Over-production of atrial natriuretic factor	Chronic hypotension
Introduction of rat angiotensinogen gene	Hypertension
Constitutively active tyrosine kinase	Cardiac hypertrophy
Over-expression of amyloid precursor protein	Alzheimer's disease
Trisomy 16	Alzheimer's disease
Expression of Simian cholesteryl ester transfer protein	Atherosclerosis

Table 1.1 Human disease equivalents derived from genetic alterations in the mouse.

even in cases where no appropriate mouse model exists. For example, it is often easier to map a new gene in the mouse and predict its location in humans than to map the gene directly in humans. In some cases the predicted location may lie near a known human disease locus and subsequent studies may show that defects in the gene are in fact responsible for the disease (Coletta *et al.* 1994).

Why sequence genomes?

DNA is the genetic material in all cells. As such, it governs every facet of their existence. The information carried in the DNA determines when and how and where cells grow and divide: for example, why some yeast cells divide by budding and others by binary fission. But what are the differences at the DNA level? And why are some cells prokaryotic and others eukaryotic and what is the difference at the DNA level between simple and complex eukaryotes? DNA has provided the basis for the evolutionary process that has generated the millions of different life-forms that exist on earth. We are all familiar with the story of Darwin's finches, but what happened at the level of the DNA? The answers to these questions, and many others like them, can be inferred from the vast knowledge we have of cellular and molecular biochemistry. However, they can only be answered definitively by studying the sequence of DNA in different organisms.

Although much of our knowledge of cellular and molecular biology was elucidated without DNA sequence information, this was cause rather than effect. Techniques for sequencing DNA did not exist.

However, a complete understanding of many biological phenomena was dependent on the development of DNA sequencing techniques. Furthermore, recent experience has shown that the analysis of sequence data is a cost-effective way to generate answers to fundamental questions like those raised above, i.e. sequencing at the beginning of an investigation can be just as worth while, if not more so, than at the end. But, is it necessary to sequence entire genomes? The answer to this must be in the positive.

Detailed understanding of an organism will only be achieved when every gene has been identified and its transcript and the timing of transcript synthesis known. As a minimum this demands knowledge of the complete gene sequence. In this context it is worth noting that when this was first available for a chromosome, that of yeast chromosome III (Oliver *et al.* 1992), the gene density was much higher than expected. An understanding of evolution will require comparative analysis of entire genomes rather than individual genes. For example, many bacterial genes are organized into operons which are transcribed as polycis-tronic mRNAs. By contrast, eukaryotic genes were thought to be regulated individually and transcribed as monocistronic mRNAs. Now analysis of 2 megabases (Mb) of DNA sequence from the nematode *Caenorhabditis elegans* has shown that it too has operons, i.e. it uses both the prokaryotic and eukaryotic patterns of gene organization (Zorio *et al.* 1994). Is the *C. elegans* genome becoming more compact to achieve bacterial status or is it expanding towards the eukaryotic monocistronic design?

Although the mapping of disease genes in humans, and useful traits in plants and animals, has been undertaken successfully in the past with an absence of sequence information, it is now generally recognized that a genome-wide effort will be more efficient in the long run. This is particularly true when one is seeking to control complex or quantitative traits (see Chapter 6) or to separate the genetic and environmental components of particular diseases. Pharmaceutical companies have taken a great interest in the human genome project for they see the data generated providing an understanding of multifactorial disease. This will enable them to design drugs that will treat the *causes* of disease rather than the *symptoms*, e.g. drugs that act at the level of transcription as opposed to the protein product or, alternatively, oligonucleotides for gene therapy. At the end of the day, the most detailed map available would be one in which every base pair had been identified.

Genome sequencing projects

A large number of different genomes are being sequenced and the most important are listed in Table 1.2. Each has been selected for a different reason. There should be no surprise over the choice of *Escherichia coli*. Of all organisms it probably is the best characterized,

Table 1.2 Key organisms whose genomes are being sequenced.

Escherichia coli
Bacillus subtilis
Saccharomyces cerevisiae (baker's yeast)
Schizosaccharomyces pombe (fission yeast)
Caenorhabditis elegans (nematode)
Drosophila melanogaster (fruitfly)
Arabidopsis thaliana (thale cress)
Oryza sativa (rice)
Mus musculus (mouse)
Homo sapiens (man)

both genetically and biochemically. *Bacillus subtilis* is of interest, partly because it is Gram positive whereas *E. coli* is Gram negative, and partly because it undergoes differentiation during the process of sporulation. *Saccharomyces cerevisiae* can be handled like a microorganism but is eukaryotic. Because it has a nucleus and chromosomes, undergoes meiosis and mitosis, etc. analysis of its genome sequence should provide much information on what constitutes a eukaryote. By the end of 1994 the sequences of four complete chromosomes had been reported (Feldmann *et al.* 1994) and *S. cerevisiae* is on track to be the first cellular organism whose entire genome has been sequenced. Unlike most cells, *S. cerevisiae* multiplies by budding, hence, the interest in *Schizosaccharomyces pombe* which divides by binary fission. Another reason for interest in *S. pombe* is that it has a genome of similar size to that of *S. cerevisiae* but is poorly understood at the genetic level. Thus it can be a useful test system for new mapping and sequencing methods. *Caenorhabditis elegans* is a nematode worm and, as such, is a simple multicellular organism. This organism has been extensively analysed genetically but, more important, the line of descent from the zygote is known for every one of the 2000 cells in the adult worms. In this organism it is possible to identify which genes are expressed and when and in what cells as the different cell lineages branch off in the course of determination and differentiation. The inclusion of *Drosophila melanogaster* should be no surprise. Not only is it well studied genetically, having been a favourite model for Mendelian genetics, but the genetics of morphogenesis and embryogenesis are well understood. The interest in man stems largely from a desire to understand inherited disease and to find ways of diagnosing it, treating it or preventing it.

The 5-year goals of the US human genome project have been published (Collins & Galas 1993) and give a good indication of the technological needs and mapping and sequencing objectives of the genome community as a whole. These objectives are summarized in Table 1.3 and it is worth noting that they incorporate sequencing of the genomes of model organisms, all of which have been described

Table 1.3 Goals of the 5-year plan for the US human genome project (from Collins & Galas 1993).

Genetic map

Complete the 2–5 cM human genetic map by 1995

Develop technology to allow non-expert to type families for medical research

Develop markers that can be screened in automated fashion

Develop new mapping technologies

Physical map

Complete map of the human genome at a resolution of 100 kb

Generate clone libraries with improved stability and lower chimaerism

DNA sequencing

Develop efficient approaches to sequencing several megabases of DNA

Develop technology for high throughput sequencing

Build up a sequencing capacity to a collective rate of 50 Mb per year

Gene identification

Develop efficient methods of identifying genes and for placement of known genes on physical maps or sequenced DNA

Technology development

Develop evolutionary and revolutionary new technology

Model organisms

Finish an STS map of mouse genome at 300 kb resolution

Finish the sequence of the *E. coli* and *S. cerevisiae* genomes by 1998 (or earlier)

Continue sequencing *C. elegans* and *Drosophila* genomes with the aim of completing *C. elegans* in 1998

Sequence selected segments of mouse DNA side by side with corresponding human DNA in areas of high biological interest

Informatics

Continue to create, develop and operate databases including effective tools and standards for data exchange

Continue to develop better software for mapping, sequencing, data comparison and analysis

Ethical, legal and social issues

Continue to identify and define issues and develop policy options

Develop policy options regarding genetic testing servicesm

Training

Continue to encourage training of scientists in interdisciplinary sciences related to genome research

Technology transfer

Encourage and enhance technology transfer both into and out of centres of genome research

---

above. Since directed breeding is not permitted in man for ethical reasons, the mouse (*Mus musculus*) is a useful model system. In addition, the mouse is an important species for the evaluation of pharmaceuticals, hence the need for a comparative understanding of its genome relative to man. For commercial reasons, genome maps at least are being constructed for all the major livestock species, particularly with a view to understanding quantitative traits such as size, protein vs. fat content, weight gain, etc.

A number of genome sequencing projects are under way for the major crop species, particularly cereals. Of these, rice is the most important since not only does it have the smallest cereal genome (430Mb) but it is a staple food of more than half the world's population. The principal objective of the rice genome project is the identification of genes that will provide resistance to the growing ravages of bacterial, fungal and viral diseases as well as to salinity and drought. It also serves as a model monocot plant. *Arabidopsis thaliana* increasingly is being used as a model organism for general plant genome analysis. It is a small flowering plant which is easily grown, has a rapid life cycle and gives a prolific seed set. These features, plus the fact that it is diploid and can be self- or cross-pollinated, make it attractive for classical genetics. Its small genome (150 Mb) makes it the dicot of choice for a sequencing project.

The fascination of genome sequencing and mapping

Following the development in 1977 of robust and generally applicable sequencing methodologies, good progress was made in sequencing a number of viral genomes, e.g. the 48.5 kb of phage λ (Sanger *et al.* 1982) and the 172 kb Epstein Barr virus DNA (Baer *et al.* 1984). This generated much excitement and led a decade ago to proposals to sequence the human genome (3×10^9 bp). However, these proposals engendered much controversy that encompassed politics, personalities and scientific judgements of the probable future course of developments in mapping and sequencing. Among the concerns raised were issues such as: 'is the sequencing of the human genome an intellectually appropriate project for biologists, is sequencing the human genome feasible, what benefits might arise from the project, will these benefits justify the cost and are there alternative ways of achieving the same benefits and will the project compete with other areas of biology for funding and intellectual resources?'

It has to be realized that, to be carried out effectively, genome sequencing must be managed by skilled scientists. It also is repetitive, boring and labour intensive. These two aspects are not compatible and can lead to a high staff turnover rate. However, if the task was completely uninteresting the human genome project would never have got off the ground, far less have spawned the various sequencing projects listed in Table 1.2. The answer to this paradox lies in the complexity of the task. The different genome sizes and architecture in different organisms mean that different mapping strategies are required and a whole series of methods has evolved. Many of these are very elegant but also very complex. As each new method is applied whole new avenues of research open up. Indeed, so much information has been generated that a number of dedicated journals, e.g. *Nature Genetics* and *Genomics*, have been created. Rather than being a mindless task, genome mapping and sequencing have become intellectually very demanding. For the experienced molecular or

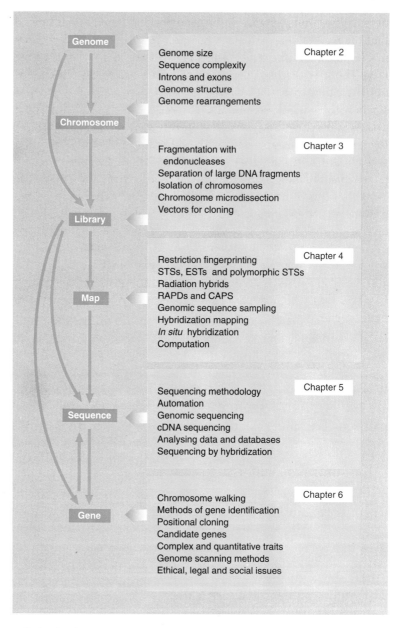

Figure 1.4 'Road map' outlining the different methodologies and interrelationships of genome mapping and sequencing covered in this book.

cellular biologist entering this field for the first time, never mind the novice, an understanding of the different methods and their interrelationships is nigh on impossible. The objective of this book is to act as a road map (Fig. 1.4). However, the reader does need some basic knowledge, principally a familiarity with hybridization in its different formats, cloning and polymerase chain reaction (PCR) technology. The reader who does not have these should first consult the monograph by Old & Primrose (1994).

Outline of the rest of the book (Fig. 1.4)

As noted above, it is necessary to have an understanding of the organization and structure of the genomes of different organisms so that the methodology used can be optimized. This topic is covered in Chapter 2 and the reader will soon realize the magnitude of the task. The sheer size of the genome of even a bacterium is such that to handle the DNA in the laboratory we need to break it down into smaller pieces which are handled as clones. The methods for doing this are covered in Chapter 3. The process of putting the pieces back together again is mapping and the different methods of mapping and the different markers used are described in Chapter 4. DNA sequencing technology currently is such that only short stretches (~600 bp) can be analysed in a single reaction. Consequently, the genome has to be fragmented and the sequence of each fragment determined and the total sequence re-assembled (Chapter 5). Fortunately, the tools and techniques used for mapping also can be applied to genome sequencing. Finally, a combination of mapping and sequencing can be used to isolate genes corresponding to traits of interest and to determine gene function. This is covered in the final chapter.

2 The organization and structure of genomes

Introduction

There is no such thing as a common genome structure. Rather, there are major differences between the genomes of bacteria, viruses and organelles on the one hand and the nuclear genomes of eukaryotes on the other. Within the eukaryotes there are major differences in the types of sequences found, the amounts of DNA and the number of chromosomes. This wide variability means that the mapping and sequencing strategies involved depend on the individual genome being studied.

Genome size

Because the different cells within a single organism can be of different ploidy, e.g. germ cells are usually haploid and somatic cells diploid, genome sizes always relate to the haploid genome. The size of the haploid genome also is known as the C-value. Measured C-values range from 3.5×10^3 bp for the smallest viruses, e.g. coliphage MS2, to 10^{11} bp for some amphibians and plants (Fig. 2.1). The largest viral genomes are $1-2 \times 10^5$ bp and are just a little smaller than the smallest cellular genomes, those of some mycoplasmas (5×10^5 bp). Simple unicellular eukaryotes have a genome size ($1-2 \times 10^7$ bp) that is not much larger than that of the largest bacterial genomes. Primitive multicellular organisms such as nematodes have a genome size about four times larger. Not surprisingly, an examination of the genome sizes of a wide range of organisms has shown that the *minimum* C-value found in a particular phylum is related to the structural and organizational complexity of the members of that phylum. Thus the minimum genome size is greater in organisms that evolutionary are more complex (Fig. 2.2).

A particularly interesting aspect of the data shown in Fig. 2.1 is the range of genome sizes found within each phylum. Within some phyla, e.g. mammals, there is only a twofold difference between the largest and smallest C-value. Within others, e.g. insects and plants, there is a 10- to 100-fold variation in size. Is there really a 100-fold variation in the number of genes needed to specify different flowering

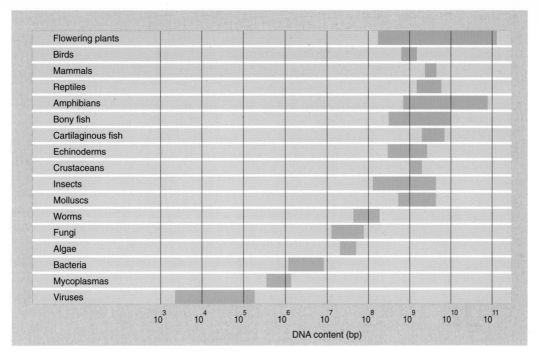

Figure 2.1 The DNA content of the haploid genome of a range of phyla. The range of values within a phylum is indicated by the shaded area. (Redrawn with permission from Lewin 1994.)

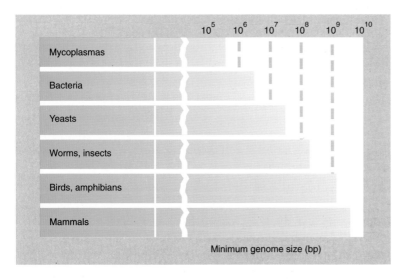

Figure 2.2 The minimum genome size found in range of organisms. (Redrawn with permission from Lewin 1994.)

plants? Are some plants really more organizationally complex than man as these data imply? The resolution of this apparent C-value paradox has been provided by the analysis of sequence complexity by means of reassociation kinetics.

Sequence complexity

When double-stranded DNA in solution is heated it denatures

('melts') releasing the complementary single strands. If the solution is cooled quickly the DNA remains in a single-stranded state. However, if the solution is cooled slowly reassociation will occur. The conditions for efficient reassociation of DNA were determined originally by Marmur *et al.* (1963) and since then have been extensively studied by others (for a review, see Tijssen 1993). The key parameters are as follows. First, there must be an adequate concentration of cations and below 0.01 M sodium ion there is effectively no reassociation. Second, the temperature of incubation must be high enough to weaken intrastrand secondary structure. In practice, the optimum temperature for reassociation is 25°C below the melting temperature (T_m), that is, the temperature required to dissociate 50% of the duplex. Third, the incubation time and the DNA concentration must be sufficient to permit an adequate number of collisions so that the DNA can reassociate. Finally, the size of the DNA fragments also affects the rate of reassociation and is conveniently controlled if the DNA is sheared to small fragment.

The reassociation of a pair of complementary sequences results from their collision and therefore the rate depends on their concentration. Since two strands are involved the process follows second-order kinetics. Thus, if C is the concentration of DNA that is single stranded at time t, then

$$\frac{dC}{dt} = -kC^2$$

where k is the reassociation rate constant. If C_0 is the initial concentration of single-stranded DNA at time $t = 0$, integrating the above equation gives

$$\frac{C}{C_0} = \frac{1}{1 + k.C_0t}.$$

When the reassociation is half complete, $C/C_0 = 0.5$ and the above equation simplifies to

$$C_0t_{1/2} = \frac{1}{k}.$$

Thus the greater the $C_0t_{1/2}$ value, the slower the reaction time at a given DNA concentration. More important, for a given DNA concentration the half-period for reassociation is proportional to the number of different types of fragments (sequences) present and thus to the genome size (Britten & Kohne 1968). This can best be seen from the data in Table 2.1. Since the rate of reassociation depends on the concentration of complementary sequences, the $C_0t_{1/2}$ for organism B will be 200 times greater than for organism A.

Experimentally it has been shown that the rate of reassociation is indeed dependent on genome size (Fig. 2.3). However, this proportionality is only true in the absence of repeated sequences. When the reassociation of calf thymus DNA was first studied kinetic analysis

CHAPTER 2
*The organization and
structure of genomes*

indicated the presence of two components (Fig. 2.4). About 40% of the DNA had a $C_0t_{1/2}$ of 0.03 whereas the remaining 60% had a $C_0t_{1/2}$ of 3000. Thus the concentration of DNA sequences that reassociate rapidly is 100 000 times the concentration of those sequences that reassociate slowly. If the slow fraction is made up of unique sequences, each of which occurs only once in the calf genome, then the sequences of the rapid fraction must be repeated 100 000 times on average. Thus the $C_0t_{1/2}$ value can be used to determine the sequence complexity of a DNA preparation. A comparative analysis of DNA from different sources has shown that repetitive DNA occurs widely in eukaryotes (Davidson & Britten 1973) and that different types of repeat are present. In the example shown in Fig. 2.5 a fast-renaturing and an intermediate-renaturing component can be recognized and are present in different copy numbers (500 000 and 350, respectively) relative to the slow component which is unique or non-repetitive DNA. The complexities of each of these components are 340 bp, 6×10^5 bp and 3×10^8 bp, respectively. The proportion of the genome that is occupied by non-repetitive DNA versus repetitive DNA varies in

	Organism A	Organism B
Starting DNA concentration (C_0)	10 pg/ml	10 pg/ml
Genome size	0.01 pg	2 pg
No. of copies of genome per ml	1000	5
Relative concentration (A vs. B)	200	1

Table 2.1 Comparison of sequence copy number for two organisms with different genome sizes.

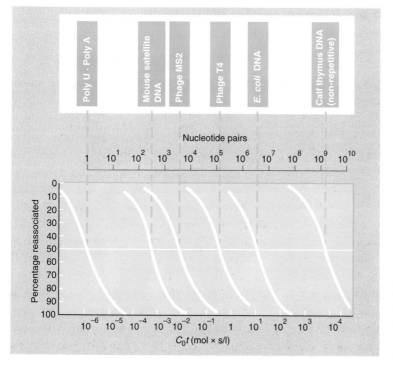

Figure 2.3 Reassociation of double-stranded nucleic acids from various sources. (Redrawn with permission from Lewin 1994.)

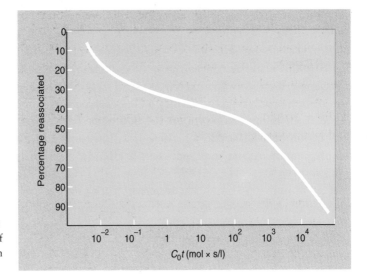

Figure 2.4 The kinetics of reassociation of calf thymus DNA. Compare the shape of the curve with that shown in Fig. 2.3.

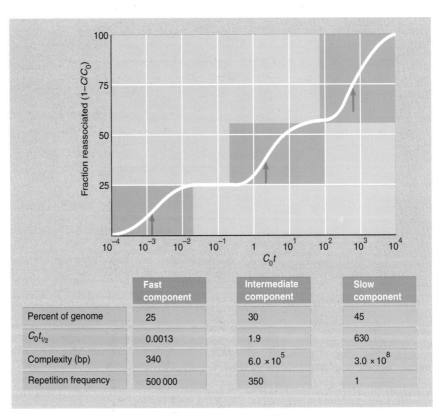

	Fast component	Intermediate component	Slow component
Percent of genome	25	30	45
$C_0t_{1/2}$	0.0013	1.9	630
Complexity (bp)	340	6.0×10^5	3.0×10^8
Repetition frequency	500 000	350	1

Figure 2.5 The reassociation kinetics of a eukaryotic DNA sample showing the presence of two types of repeated DNA. The arrows indicate the $C_0t_{1/2}$ values for the three components. (Redrawn with permission from Lewin 1994.)

different organisms (Fig. 2.6) thus resolving the C-value paradox. In general, the length of the non-repetitive DNA component tends to increase as we go up the evolutionary tree to a maximum of 2×10^9 bp in mammals. The fact that many plants and animals have a much

CHAPTER 2
*The organization and
structure of genomes*

higher C-value is a reflection of the presence of large amounts of repetitive DNA. Analysis of mRNA hybridization to DNA shows that most of it anneals to non-repetitive DNA, i.e. most genes are present in non-repetitive DNA. Thus genetic complexity is proportional to the content of non-repetitive DNA and not to genome size. There is, however, a surprising constancy in gene numbers. In all eukaryotes there are 7000–20 000 genes, except for the vertebrates which have 50 000–100 000 genes (Bird 1995).

Introns and exons

Introns were initially discovered in the chicken ovalbumin and rabbit and mouse β-globin genes (Breatnach *et al.* 1977; Jeffreys & Flavell 1977). Both these genes had been cloned by isolating the mRNA from expressing cells and converting it to cDNA. The next step was to use the cloned cDNA to investigate possible differences in the structure of the gene from expressing and non-expressing cells. Here the Southern blot hybridizations revealed a totally unanticipated situation. It was expected that the analysis of genomic restriction fragments generated by enzymes that did not cut the cDNA would reveal only a single band corresponding to the entire gene. Instead several bands were detected in the hybridized blots. The data could be explained only by assuming the existence of interruptions in the middle of the protein-coding sequences. Furthermore, these insertions

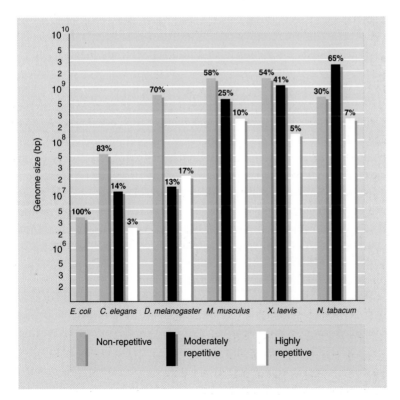

Figure 2.6 The proportions of different sequence components in representative eukaryotic genomes. (Redrawn with permission from Lewin 1994.)

appeared to be present in both expressing and non-expressing cells. The gene insertions that are not translated into protein were termed *introns* and the sequences that are translated were called *exons*.

Since the original discovery of introns a large number of split genes have been identified in a wide variety of organisms. These introns are not restricted to protein-coding genes for they have been found in rRNA and tRNA genes as well. Split genes are rare in prokaryotes and not particularly common in lower eukaryotes such as yeast. However, proceeding up the evolutionary tree, the number of split genes, and the number and size of introns per gene, increases (Fig. 2.7 and Table 2.2). More important, genes that are related by evolution have exons of similar size, i.e. the introns are in the same position. However, the introns may vary in length giving rise to variation in the length of the genes (Fig. 2.8). Note also that introns are much longer than exons, particularly in higher eukaryotes.

If a split gene has been cloned it is possible to sub-clone either the exon or the intron sequences. If these sub-clones are used as probes in genomic Southern blots it is possible to determine if these same sequences are present elsewhere in the genome. Often, the exon sequences of one gene are found to be related to sequences in one or more other genes. Some examples of such *gene families* are given in Table 2.3. In some instances the duplicated genes are clustered whereas in others they are dispersed. Also, the members may have related or even identical functions although they may be expressed at different times or in different cell types. Thus different globin proteins are found in embryonic and adult red blood cells while different actins are found in muscle and non-muscle cells. Functional divergence between members of a multigene family may extend to the loss of gene function by some members. Such *pseudogenes* come in two types. In the first type they retain the usual intron and exon structure but are functionless or they lack one or more exons. In the second type, found in dispersed gene families, processed pseudogenes are found which lack any sequences corresponding to the introns or promoters of the functional gene members. Multiple copies of an exon also may be found because the same exons occur in several apparently unrelated genes. Exons that are shared by several genes are likely to encode polypeptide regions that endow the disparate proteins with related properties, e.g. ATP or DNA binding. Some genes appear to be mosaics that were constructed by patching together copies of individual exons recruited from different genes, a phenomenon known as *exon shuffling*.

By contrast with exons, introns are not related to other sequences in the genome although they contain the majority of dispersed, highly repetitive sequences. Thus, for some genes the exons constitute slightly repetitive sequences embedded in a unique context of introns.

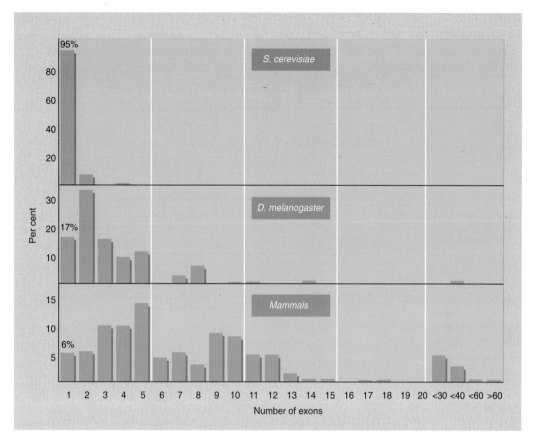

Figure 2.7 The number of exons in three representative eukaryotes. Uninterrupted genes have only one exon and are totalled in the left hand column. (Redrawn with permission from Lewin 1994.)

Species	Average exon number	Average intron number	Average length (kB)	Average mRNA length (kB)	% Exon per gene
Yeast	1	0	1.6	1.6	100
Nematode	4	3	4.0	3.0	75
Fruitfly	4	3	11.3	2.7	24
Chicken	9	8	13.9	2.4	17
Mammals	7	6	16.6	2.2	13

Table 2.2 Intron statistics for genes from different animal species.

Genome structure in viruses and prokaryotes

The genomes of viruses and prokaryotes are very simple structures although those of viruses show remarkable diversity (for review see Dimmock & Primrose 1994). Most viruses have a single linear or circular genome but a few such as the reoviruses, bacteriophage Φ6 and some plant viruses have segmented RNA genomes. Until relatively recently it was believed that all eubacterial genomes consisted of a single circular chromosome. However, linear

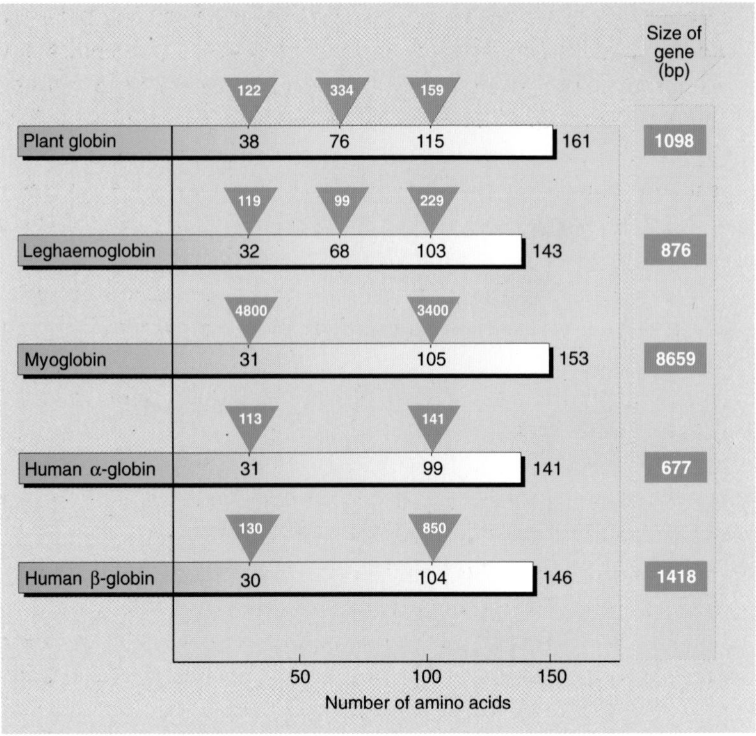

Figure 2.8 The placement of introns in different members of the globin superfamily. The size of the introns in base pairs is indicated inside the inverted triangles. Note that the size of each polypeptide and the location of the different introns are relatively consistent.

Table 2.3 Some examples of multigene families.

Gene family	Organism	Approximate no. of genes	Clustered (L) or dispersed (D)
Actin	Yeast	1	–
	Slime mould	17	L, D
	Drosophila	6	D
	Chicken	8–10	D
	Human	20–30	D
Tubulin	Yeast	3	D
	Trypanosome	30	L
	Sea urchin	15	L, D
	Mammals	25	D
α-Amylase	Mouse	3	L
	Rat	9	?
	Barley	7	?
β-Globin	Human	6	L
	Lemur	4	L
	Mouse	7	L
	Chicken	4	L

CHAPTER 2
*The organization and
structure of genomes*

chromosomes have been found in *Borrelia burgdorfii*, *Streptomyces lividans* and *Rhodococcus fascians* while two chromosomes have been found in *Rhodobacter spheroides*, *Brucella melitensis*, *Leptospira interrogans* and *Agrobacterium tumefaciens* (Cole & Saint Girons 1994). Where two chromosomes have been detected, essential genes for growth are found on both. Linear plasmids have been found in *Borrelia* spp. and *Streptomyces* spp. (Kinashi *et al.* 1987; Barbour 1993).

Bacterial and viral genomes lack the centromeres found in eukaryotic chromosomes although there may be a partitioning system based on membrane adherence. Duplication of the genomes is initiated at an origin of replication and may proceed unidirectionally or bidirectionally. The structure of the origin of replication, the *ori*C locus, has been extensively studied in a range of bacteria and found to consist essentially of the same group of genes in a nearly identical order (Cole & Saint Girons 1994). The *ori*C locus is defined as a region harbouring the *dna*A (DNA initiation) or *gyr*B (B subunit of DNA gyrase) genes linked to a ribosomal RNA operon.

Many bacterial and viral genomes are either circular or can adopt a circular conformation for the purposes of replication. As such, these molecules do not have telomeres. However, many viral genomes always retain a linear configuration, as do linear plasmid molecules. All these molecules exhibit specialized structural features to facilitate duplication of the ends of the molecules and to protect them from exonuclease digestion. For example, poxviruses have closed hairpin loops at the end of their genomes and phages T2 and T7 exhibit terminal redundancy, i.e. they have the sequence at both ends of the viral DNA. Many other variations have been noted (for review, see Dimmock & Primrose 1994). The structure of the chromosome ends in bacteria with linear chromosomes is not known.

In the original studies on the kinetic analysis of the reassociation of phage and bacterial DNA no repeated sequences were noted (see p. 16). As a further check for repetitive DNA in *E. coli*, Britten and Kohne (1968) isolated the first small fraction to reassociate and compared its reassociation kinetics with that of bulk *E. coli* DNA. No differences were found. While the sensitivity of this test is high, the existence of a small amount of repetitive DNA cannot be ruled out. Indeed, repetitious DNA has been detected. For example, classical genetic studies have revealed the presence of multiple copies of mobile elements known as insertion sequences in the genome of *E. coli* and its relatives. However, the copy number is very small, usually less than 20 per genome, and they represent a very small proportion of the total cellular DNA. Sequencing of the *E. coli* genome has revealed the presence of other repeated sequences such as REP (repeated extragenic palindrome) and ERIC (enterobacterial repetitive intergenic consensus) elements. For example, a 225 kb region of the *E. coli* genome was found to contain 19 REP elements

(Sofia *et al.* 1994). Also in *E. coli*, a large number of pairs or groups of redundant genes have been identified (Riley 1993) in addition to multiple genes for tRNAs. These redundant genes have similar nucleotide sequences and the gene products similar amino-acid compositions.

Split genes appear to be rare in prokaryotes and viruses. Introns have been detected in only a few bacteriophages and in a tRNA[ser] gene in the archaebacterium *Acanthamoeba* (Belfort 1989).

The organization of organelle genomes

Mitochondria and chloroplasts both possess DNA genomes that code for all of the RNA species and some of the proteins involved in the functions of the organelle. In some lower eukaryotes the mitochondrial (mt) DNA is linear but more usually organelle genomes take the form of a single circular molecule of DNA. Because each organelle contains several copies of the genome and because there are multiple organelles per cell, organelle DNA constitutes a repetitive sequence. Whereas chloroplast (ct) DNA falls in the range 120–200 kb, mt DNA varies enormously in size. In animals it is relatively small, usually less than 20 kb. But yeast mt DNA is about 80 kb and that of plants ranges from several hundred to several thousand kilobase pairs. In addition, certain protozoans, e.g. trypanosomes, have within their single mitochondrion a disk structure known as a kinetoplast. This contains thousands of interlocked circular DNA molecules. In *Trypanosoma brucei* about 45 of the circles are 21 kb long and the remaining 5500 are 1 kb long. Considerable sequence variation occurs within the circles (Lamond 1988).

ORGANIZATION OF THE CHLOROPLAST GENOME

The complete sequence of ct DNA has been reported for a liverwort (a bryophyte; Ohyama *et al.* 1986), tobacco (a dicot; Shinozaki *et al.* 1986) and rice (a monocot; Hiratsuka *et al.* 1989) and the general organization of many other ct DNA molecules has been elucidated by restriction endonuclease mapping (see later). A general feature of ct DNA from both higher and lower eukaryotes is a 10–24 kb sequence that is present in two identical copies as an inverted repeat (Fig. 2.9). The only exceptions to this structure are in some legumes that have lost one copy of the inverted repeat and *Euglena* which lacks both copies. However, *Euglena* does have three tandem repeats of a 5.6 kb sequence. Key features of the ct DNA genome are shown in Table 2.4.

ORGANIZATION OF THE MITOCHONDRIAL GENOME

Because mt DNA varies so widely in size it is difficult to make

CHAPTER 2
*The organization and
structure of genomes*

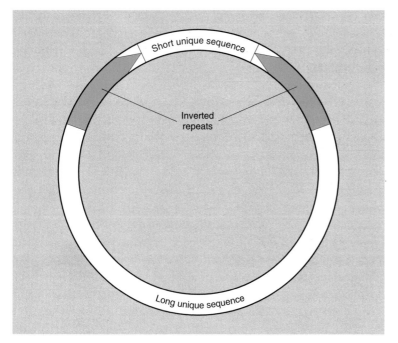

Fig. 2.9 Generalized
structure of ct DNA.

Table 2.4 Key features of ct
DNA.

Feature	Liverwort	Tobacco	Rice
Inverted repeats	10 058 bp	25 339 bp	20 799 bp
Short unique sequence	19 813 bp	18 482 bp	
Long unique sequence	81 095 bp	86 684 bp	
Length of total genome	121 024 bp	155 844 bp	134 525 bp
Number of genes	128	84	
Number of genes with introns	20	15	18

generalizations about its organization. The 16.6 kb mt DNA from three species (man, mouse and cow) have been sequenced and the genome organization is similar: there are no introns, some genes overlap and almost every single base pair can be assigned to a gene. The mt DNA from *S. cerevisiae* is 84 kb long, and has not been sequenced. However, about 24% of the genome is known to consist of short stretches of A–T-rich DNA and introns are known to occur.

The organization of nuclear DNA in eukaryotes

GROSS ANATOMY

Each eukaryotic nucleus encloses a fixed number of chromosomes which contain the nuclear DNA. During most of a cell's life, its chromosomes exist in a highly extended linear form. Prior to cell division, however, they condense into much more compact bodies which can be examined microscopically after staining. The duplication of chromosomes occurs chiefly when they are in the extended stage (interphase).

One part of the chromosome, however, always duplicates during the contracted metaphase state: this is the *centromere*, a body that controls the movement of the chromosome during mitosis. Its structure is discussed later (p. 34).

In many eukaryotes, a variety of treatments will cause chromosomes in dividing cells to appear as a series of light and dark staining bands (Fig. 2.10). In G-banding, for example, the chromosomes are subjected to controlled digestion with trypsin before Giemsa staining which reveals alternating positively (dark G-bands) and negatively (pale G-bands) staining regions. As many as 2000 light and dark bands can be seen along some mammalian chromosomes. An identical banding pattern (Q-banding) can be seen if the Giemsa stain is replaced with a fluorescent dye such as quinacrine which intercalates between the bases of DNA. Bands are classified according to their relative location on the short arm (p) or the long arm (q) of specific chromosomes. For example, 12q1 means band 1 on the long arm of chromosome 12. If the chromosome DNA is treated with acid and then alkali prior to Giemsa staining then only the centromeric region stains and this is referred to as *heterochromatin*. The unstained parts of the chromosome are called *euchromatin*.

Because Giemsa shows preferential binding to DNA rich in AT base pairs, the dark G-bands are believed to be A+T rich and light G-bands G+C rich. The G+C-rich regions are believed to contain the majority of genes. It should be noted that DNA which is not transcriptionally active tends to be methylated. Because methylation of restriction enzyme recognition sequences can prevent endonuclease cleavage this can affect the recovery of DNA fragments.

The ends of eukaryotic chromosomes are also the ends of linear duplex DNA and are known as *telomeres*. That these must have a special structure has been known for a long time. For example, if breaks in DNA duplexes are not rapidly repaired by ligation they undergo recombination or exonuclease digestion, Yet, the ends of

Figure 2.10 Banding patterns revealed on chromosomes by different staining methods. Note that intercalating flourescent dyes produce the same pattern as Giemsa stain at 60°C. In the C-banding technique some heterochromatin may be detected at the telomeres.

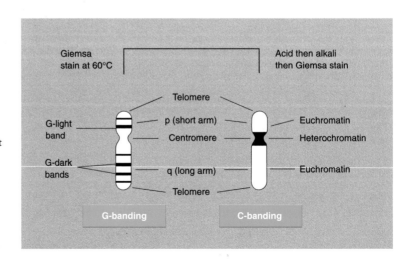

CHAPTER 2
*The organization and
structure of genomes*

chromosomes are stable and chromosomes do not get ligated together. Also, DNA replication is initiated in a 5'→3' direction with the aid of an RNA primer. After removal of this primer there is no way of completing the 5' end of the molecule (Fig. 2.11). Thus, in the absence of a method for completing the ends of the molecules, chromosomes would become shorter after each cell division. As will be seen later, the ends of chromosomes have an unusual structural organization.

ORIGINS OF REPLICATION

All of the cell's DNA must be replicated once, and once only, during the cell cycle. In yeast (*S. cerevisiae*) the origins of DNA replication have been identified as ARS (autonomously replicating sequence) elements. Incorporation of the latter on plasmids permits their replication in yeast as yeast artificial chromosomes (YACs; see p. 49). The sequence requirements for an ARS element are known (Marahrens & Stillman 1992). So far, origins of DNA replication have not been unambiguously identified in other eukaryotes.

REPEATED SEQUENCES

It will be recalled from the section on reassociation kinetics of DNA that repeated sequences are a common feature of eukaryotic DNA. How are these repeated sequences organized in the genome? Are they tandemly repeated or are they dispersed? The answer is that both organizational patterns are found. In general, highly repetitive DNA is organized around centromeres and telomeres in the form of tandem repeats whereas moderately repetitive DNA is dispersed throughout the chromosome.

Tandemly repeated sequences

Repeated sequences were first discovered 30 years ago during studies on the behaviour of DNA in centrifugal fields. When DNA is centrifuged to equilibrium in solutions of CsCl, it forms a band at the position corresponding to its own buoyant density. This in turn

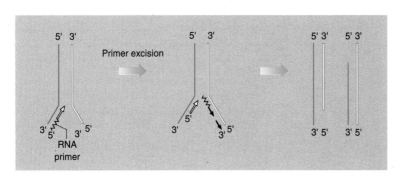

Figure 2.11 Formation of two daughter molecules with complementary single-stranded 3' tails after primer excision.

27

depends on its percentage G + C content:

$$\rho\text{(density)} = 1.660 + 0.00098 \ (\% \ GC) \ \text{g/cm}^3.$$

When eukaryotic DNA is centrifuged in this way the bulk of the DNA forms a single rather broad band centred on the buoyant density which corresponds to the average G:C content of the genome. Frequently one or more minor or *satellite* bands are seen (Fig. 2.12). The behaviour of satellite DNA on density gradient centrifugation frequently is anomalous. When the base composition of a satellite is determined by chemical means it often is different from what had been predicted from its buoyant density. One reason is that it is methylated which changes its buoyant density.

Once isolated, satellite DNA can be radioactively labelled *in vitro* and used as a probe to determine where on the chromosome it will hybridize. In this technique, known as *in situ hybridization*, the chromosomal DNA is denatured by treating cells that have been squashed on a cover slip. The localization of the sites of hybridization are determined by autoradiography. Using this technique most of the labelled satellite DNA is found to hybridize to the hetero-chromatin present around the centromeres and telomeres. Since RNA that is homologous to satellites is found only rarely the heterochromatic DNA most probably is non-coding.

When satellite DNA is subjected to restriction endonuclease digestion only one or a few distinct low-molecular-weight bands are observed following electrophoresis. These distinct bands are a tell-tale sign of tandemly repeated sequences. The reason (Fig. 2.13) is that if a site for a particular restriction endonuclease occurs in each repeat of a repetitious tandem array, then the array is digested to unit sized fragments by that enzyme. After elution of the DNA band

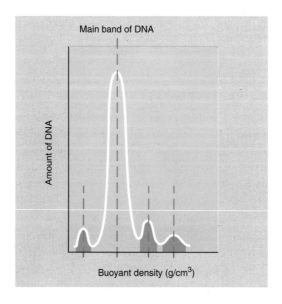

Figure 2.12 Detection of three satellite DNA bands (dark shading) on equilibrium density gradient centrifugation of total DNA.

CHAPTER 2
*The organization and
structure of genomes*

from the gel it can be used for sequence analysis either directly or after cloning. However, the sequence obtained is a consensus sequence and not necessarily the sequence of any particular repeat unit because sequence divergence can and does occur very readily. Note that if such sequence divergence occurs within a restriction endonuclease cleavage site in the repeated units then digestion with the enzyme produces multimers of the repeat unit ('higher order repeats') (Fig. 2.13).

The amount of satellite DNA and its sequence varies widely between species and can be highly polymorphic within a species. Thus 1–3% of the genome of the rat (*Rattus norvegicus*) is centromeric satellite DNA whereas it is 8% in the mouse (*Mus musculus*) and greater than 23% in the cow (*Bovis domesticus*). The length of the satellite repeat unit varies from the $d(AT)_n : d(TA)_n$ structure found in the land crab (*Gecarcinus lateratis*) to the very complicated structure seen in the domestic cow (Fig. 2.14). Even within a single genus each species can have a distinctive set of satellite sequences. In general, little is known about the detailed structure of repeated DNA arrays but one family of repeated DNA that has been analysed extensively is ribosomal DNA (rDNA) (Williams & Robbins 1992). An example of rDNA repeat structure is shown in Fig. 2.15.

Not all tandem repetitions are restricted to heterochromatin: some are found dispersed throughout the genome often in the spacer region between genes. One such group are the *minisatellite* sequences, which also demonstrates intraspecies polymorphism. This polymorphism has been used for DNA fingerprinting of organisms. When cloned probes containing a minisatellite sequence are annealed with DNA blots containing restriction endonuclease digests of DNA, multiple bands hybridize. The pattern of bands varies from one individual of a species to another but is the same when DNA from several tissues of a single individual is examined. The bands are inherited in a Mendelian fashion and it is possible to identify those bands inherited from each parent (Fig. 2.16). For this reason the technique has forensic applications (see Monckton & Jeffreys 1993 and Alford & Caskey 1994 for reviews).

Microsatellite DNA families include small arrays of tandem repeats which are simple in sequence (1–4 bp) and which are interspersed throughout the genome. In mammalian cells runs of $(dA.dT)_n$ are very common and can account for 0.3% of the genome. In contrast runs of $(dG.dC)_n$ are much rarer. Runs of dinucleotides also are found. For example, in man, runs of $(dCA.dTG)_n$ and $(dCT.dAG)_n$ account for 0.5% and 0.2% of the genome respectively. Trinucleotide and tetranucleotide repeats also occur but are rarer.

Repeated sequences also are found in telomeres and those from lower and higher eukaryotes are constructed on the same basic principle. Each telomere consists of a long series of short, tandemly repeated sequences (Table 2.5). All can be written in the general form $C_n(A/T)_m$, where $n > 1$ and m is 1–4. The number of repeats

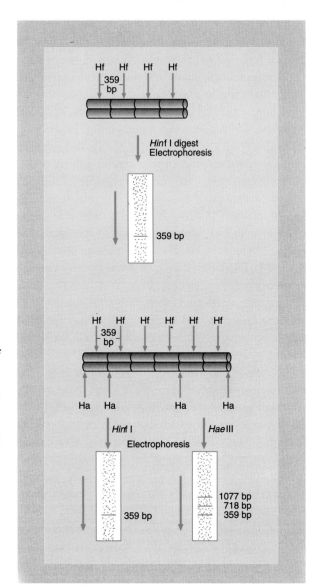

Figure 2.13 Digestion of purified satellite by restriction endonucleases. The basic repeat unit is 359 bp long and contains one endonuclease *Hin*fI (Hf) site. Digestion with *Hin*fI converts most of the satellite DNA to a set of 359 bp long fragments. These are abundant enough to be seen as a band against the smear of other genomic fragments after gel electrophoresis and staining with ethidium bromide. Digestion of the DNA with endonuclease *Hae*III (Ha) yields a ladder of fragments that are multiples of 359 bp in length. (Redrawn with permission from Singer & Berg 1990).

varies greatly but the overall telomere length can be up to 12 kb. The detailed structure of telomeres has been reviewed by Shippen (1993) and Kipling (1995).

Tandem repetition of DNA sequences also occurs within coding regions. For example, linked groups of identical or near-identical genes sometimes are repeated in tandem. These are the gene families described earlier (p. 20). However, tandem repetition also occurs within a single gene. For example, the *Drosophila* 'glue' protein gene contains 19 direct tandem repeats of a sequence 21 base pairs long that encodes seven amino acids. The repeats are not perfect but show divergence from a consensus sequence. Another example is the gene for α2(l) collagen found in chicken, mouse and man. The gene comprises 52 exons with introns varying in length from 80 to 2000

CHAPTER 2
*The organization and
structure of genomes*

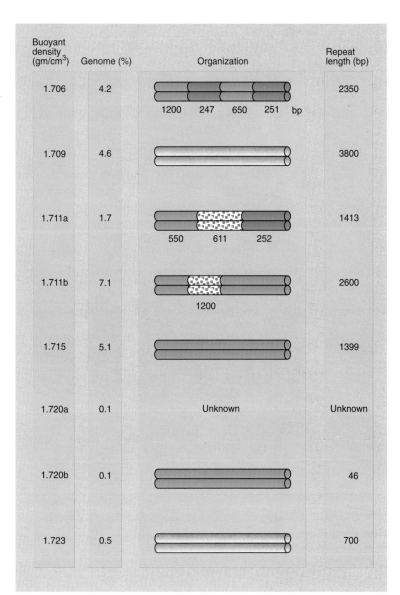

Buoyant density (gm/cm³)	Genome (%)	Organization	Repeat length (bp)
1.706	4.2	1200 247 650 251 bp	2350
1.709	4.6		3800
1.711a	1.7	550 611 252	1413
1.711b	7.1	1200	2600
1.715	5.1		1399
1.720a	0.1	Unknown	Unknown
1.720b	0.1		46
1.723	0.5		700

Figure 2.14 The structure of the different types of satellite DNA found in the domestic cow. Homologous portions of the different satellites are indicated by similar colouring. The designations a and b indicate two versions with the same buoyant density. (Redrawn with permission from Singer & Berg 1990.)

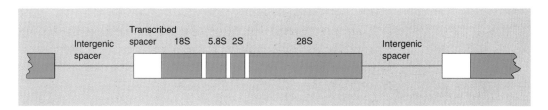

Figure 2.15 Detailed architecture of a rDNA repeat unit. Transcribed regions are shown in boxes. Coding sequences are shown in green and spacer regions in white.

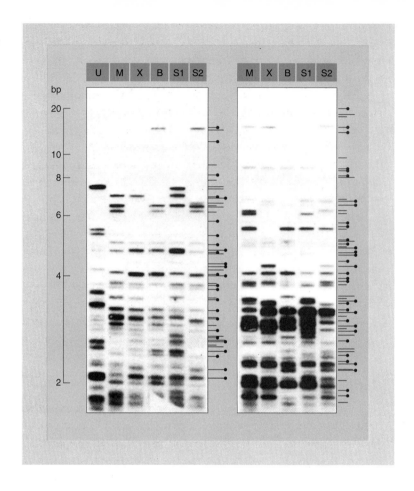

Figure 2.16 Use of minisatellite sequences to detect polymorphisms in human DNA for forensic purposes.

Table 2.5 Representative example of the repeating units found in telomeres.

Organism	Repeating unit ($5' \rightarrow 3'$)
Tetrahymena (ciliate)	CCCCAA
Trypanosoma (flagellate)	CCCTA
Dictyostelium (slime mould)	CCCTA
Saccharomyces (yeast)	$C_{2-3}A(CA)_{1-3}$
Arabidopsis (thale cress)	C_3TA_3
Homo sapiens (man)	C_3TA_2

bp. However, all the exon sequences are multiples of nine base pairs and most of them are 54 or 108 bp long. This accounts for the observed primary sequence of collagen which has glycine in every third position and a very high content of proline and lysine.

Dispersed repeated sequences

Moderately repetitive DNA is characterized by being dispersed throughout the genome and this was discovered by an extension of the early work on reannealing kinetics. It will be recalled from p. 16 that reannealing of DNA is dependent on DNA fragment size and

CHAPTER 2
*The organization and
structure of genomes*

that it is common practice to shear the DNA before hybridization begins. The reason for this is that it was observed that most high-molecular-weight fragments would reassociate with each other as a result of the interaction of repetitive DNA sequences. Very large particles, termed networks, were formed, an observation indicating that many fragments as long as 10 000 nucleotides contained more than one repetitive sequence element. When the fragment size was reduced, a smaller number of fragments reassociated at low $C_0t_{1/2}$ values. When the percentage reassociation is determined for different lengths of DNA at C_0t values where single-copy DNA cannot anneal, it is possible to determine the length of the interspersed sequences (Fig. 2.17). From the example shown it can be seen that in human DNA 2.2 kb of relatively rare sequences separate the repeated sequences in 60% of the genome. In *Xenopus* about 800 bp of unique sequences separate repeats in about 65% of the genome. By contrast, in *Drosophila* about 12 kbp separate the repeated sequences. Kinetic analysis of this type has revealed that the predominant interspersion pattern of insects and fungi is long range (i.e. like *Drosophila*), whereas in most other plants and animals it is short range. However, molecular cloning has revealed that many organisms have both short- and long-range interspersion patterns superimposed upon one another.

The short and long interspersed repeated sequences are examples of *retroposons* (Grandbastien 1992; McDonald 1993). The retro-viruses are the paradigm for retroposons that have the capacity to transpose because they code for reverse transcriptase and/or integrase activities. The retroposons differ from the retroviruses themselves in not passing through an independent infectious form but otherwise resemble them in the mechanism used for transposition. This group

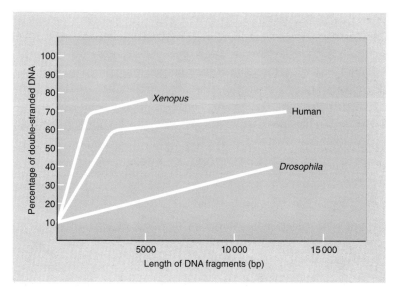

Figure 2.17 Interspersion analysis of DNA reassociation kinetics. See text for details.

CHAPTER 2
*The organization and
structure of genomes*

is called the *viral superfamily* of retroposons and examples are the
Ty element of yeast, the *copia* element of *Drosophila* and the
mammalian LINEs (long interspersed nuclear elements). These
retroposons are characterized by the presence of two open reading
frames (ORFs) on one strand, one of which encodes reverse
transcriptase. The viral superfamily of retroposons can be divided
into two groups: Those with long terminal repeats, e.g. Ty1, and
those without long terminal repeats, e.g. LINEs (Fig. 2.18).

Another group of retroposons is the *non-viral superfamily* which
does not code for proteins that have transposition functions. Rather,
they have features that suggest they originated in RNA sequences.
The best known examples are the mammalian SINEs (short, interspersed nuclear elements), which appear to be processed pseudogenes
(see p. 20) derived from genes that encode small cytoplasmic RNA
including tRNAs and 7SL RNA. The best characterized SINE family
is the Alu family found in Old World primates and named for an
*Alu*I restriction endonuclease site typical of the sequence. Alu units
are found in nearly a million copies per haploid genome and can be
found flanking genes, in introns, within satellite DNA, and in clusters
with other interspersed repeated sequences. They have an observed
average spacing distance of 3 kb but there are local regions of prefer-
ence or exclusion (Moyzis *et al.* 1989). Within the human genome,
LINEs are preferentially located in the dark G-bands of metaphase
chromosomes while SINEs are preferentially found in the light G-
bands.

CENTROMERE STRUCTURE

The structure of centromeric DNA differs widely in different
organisms. In *S. cerevisiae* it is only 170 bp long whereas in *S. pombe*
it is much longer (tens of kilobases). Centromeric DNA appears to
be very complex, at least in mammals where it has been extensively
studied (Tyler-Smith & Willard 1993). Mammalian centromeres are
made up of very large regions of repeated sequences. No unique
sequence has been found in a mammalian centromere. The most
extensively studied repeated-sequence families are the tandemly
repeated satellite DNAs but many non-satellite centromeric repeated–
sequence families also are present. The detail for the two human
centromeres is shown in Fig. 2.19.

SIGNIFICANCE OF REPEAT SEQUENCES FOR GENOME
MAPPING AND SEQUENCING

Although advantage can sometimes be taken of repeat sequences,
generally speaking they tend to cause problems. For example, when
DNA containing repeated sequences is used as a hybridization probe
it will anneal to many different regions of the genome. During cloning

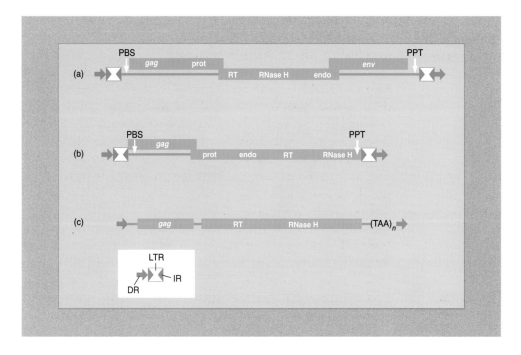

Figure 2.18 Major types of retroelements. Overall organization of (a) a retrovirus, the avian leukosis virus, (b) a retrotransposon, the yeast Ty1 element, and (c) a non-LTR (long terminal repeat) retroposon, the *Drosophila* I factor. Open reading frames are depicted by green boxes. The *gag* gene encodes structural proteins of the virion core, including a nucleic-acid-binding protein; the *env* gene encodes a structural envelope protein, necessary for cell-to-cell movement; prot, protease involved in cleavage of primary translation products; RT, reverse transcriptase; RNase H, ribonuclease; endo, endonuclease necessary for integration in the host genome; LTR, long terminal repeats containing signals for initiation and termination of transcription, and bordered by short inverted repeats (IRs) typically terminating in 5'–TG...CA–3'; PBS, primer binding site, complementary to the 3' end of a host tRNA, and used for synthesis of the first (–) DNA strand; PPT, polypurine tract used for synthesis of the second (+) DNA strand; DR, short direct repeats of the host target DNA, created upon insertion. (Redrawn with permission from Grandbastien 1992.)

of genome fragments recombination can occur between repeats leading to 'scrambling' of DNA sequences. During the actual DNA sequencing reactions, slippage can occur, particularly with microsatellites. Finally, during data assembly, incorrect positioning of genome fragments or sequences can occur because repeat units are incorrectly recognized as being unique. These issues are discussed in more detail in subsequent chapters.

SUMMARY OF CHROMOSOME STRUCTURES

The different structural elements that have been discussed are summarized in Fig. 2.20.

CHAPTER 2
*The organization and
structure of genomes*

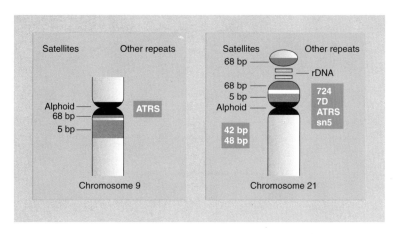

Figure 2.19 Structures of the centromeres of human chromosomes 9 and 21, showing tandemly repeated satellite DNA and other centromeric repeated sequences. The sequences shown in boxes are known to be present but their precise locations have not yet been mapped. Satellite DNA is named by the size of the repeat unit except for alphoid DNA which has a 171 bp repeat. ATRS, A+T rich sequence. (Redrawn with permission from Tyler-Smith & Willard 1993.)

Genome rearrangements

Hypotrichs are a large group of ciliated protozoa that cut, splice, reorder and eliminate DNA sequences to an extraordinary extent during their sexual life cycle (Prescott 1994). Similar kinds of DNA processing also occur in *Tetrohymena*. In hypotrichs there are two nuclei: a germ-line nucleus (micronucleus) and a somatic nucleus (macronucleus). Following cell mating a copy of the micronucleus gives rise to a new macronucleus and the old macronucleus is destroyed. All genes in the micronuclear genome are interrupted by non-functional sequences that are spliced out of the DNA to make functional macronuclear genes during development. Functional segments within some micronuclear genes are scrambled and these segments are spliced into a different order during development to yield functional macronuclear genes. A massive elimination of micronuclear sequences yields a macronucleus with a much lower sequence complexity. Finally, during processing every macronuclear gene comes to reside in a physically separate small DNA molecule which is amplified about 1000-fold.

A different kind of programmed rearrangement occurs during antibody synthesis in man. The protein sequence of an antibody is encoded by *V*, *J* and *C* genes which are physically separated on the same chromosome. There are hundreds of different *V* genes, a limited number of *J* genes and a single *C* gene for each of the heavy and light antibody chains. For an antibody to be synthesized a single *V* sequence has to be selected and joined to one of the *J* genes and the *C* gene for both the heavy and light chains. This complicated genetic

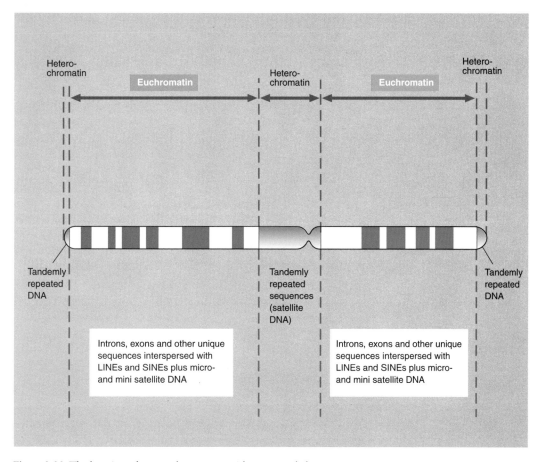

Figure 2.20 The location of repeated sequences within a typical chromosome.

arrangement gives the human body the opportunity to synthesize any of a vast repertoire of antibodies while minimizing the number of genes involved.

3 Subdividing the genome

Introduction

As outlined in Chapter 1, the first step in sequencing a genome is to divide the individual chromosomes in an ordered manner into smaller and smaller pieces that ultimately can be sequenced. That is, one begins by creating a genomic library. However, at some stage the different clones have to be ordered into a physical map corresponding to that found in the intact organism. The magnitude of this task depends on the average size of the cloned insert in the library: the larger the insert, the fewer clones that have to be ordered.

The number of clones required can be calculated easily. If the average size of the cloned insert is 20 kb and the genome has a size of 2.8×10^6 kb, e.g. the human genome, then the size of the genome relative to the size of the cloned insert, designated n, is $2.8 \times 10^6/20 = 1/4 \times 10^5$. The number of independent recombinants required in the library must be greater than n. This is because sampling variation will lead to the inclusion of some sequences several times and the exclusion of other sequences in a library of just n recombinants. Clarke & Carbon (1976) have derived a formula that relates the probability (P) of including any DNA sequence in a random library of N independent recombinants:

$$N = \frac{\ln (1 - P)}{\ln (1 - 1/n)} \, .$$

Therefore, to achieve a 95% probability ($P = 0.95$) of including any particular sequence in a random human genomic library of 20 kb fragment size:

$$N = \frac{\ln(1 - 0.95)}{\ln(1 - 1/1.4 \times 10^5)} = 4.2 \times 10^5.$$

If the probability is to be increased to 99% then N become 6.5×10^5. Put a different way, a threefold coverage gives a 95% probability of including any sequence and a fivefold coverage a 99% probability. These calculations assume equal representation of sequences, which is not true in practice.

Fragmentation of DNA with restriction enzymes

Type II restriction endonucleases have target sites which are 4–8 base pairs in length. If all bases are equally frequent in a DNA molecule then we would expect a tetranucleotide to occur on average every 4^4 (i.e. 256) nucleotide pairs in a long random DNA sequence. Similarly, a hexanucleotide would occur every 4^6 (i.e. 4096) base pairs and an octanucleotide every 4^8 (i.e. 65 536) base pairs. Table 3.1 shows the *expected* number of fragments that would be produced when different genomes are digested completely with different restriction endonucleases. In practice, the actual number of fragments is quite different because the distribution of nucleotides is non-random and most organisms do not have an equal number of the four bases. For example, human DNA has overall only 40% G+C and the frequency of the dinucleotide CpG is only 20% of that expected. Thus the enzyme *Not*I, which recognizes the sequence GCGGCCGC, cuts human DNA into fragments of average size 1000–1500 kb rather than the 65 kb expected. Similarly, the *E. coli* genome is cut by *Not*I into only 20 fragments, not the 72 expected from Table 3.1 (Smith *et al.* 1987). Again, the *Schizosaccharomyces pombe* genome, which is slightly smaller than that of *Saccharomyces cerevisiae*, is cut into only 14 fragments (Fan *et al.* 1989). In genomes rich in G+C the tetranucleotide CTAG is particularly rare (McClelland *et al.* 1987) such that the enzymes *Spe*I (ACTAGT), *Xba*I (TCTAGA), *Nhe*I (GCTAGC) and *Bln*I (CCTAGG) can produce a more limited number of DNA fragments than might at first be expected.

The average size of DNA fragment produced by digestion with restriction enzymes with 4- and 6-base recognition sequences is too small to be of much use for preparing gene libraries except in special circumstances (see below). Even enzymes with 8-base recognition sequences may not be of particular value because although the average size of fragment should be 65.5 kb, in the case of *S. pombe*, in practice

Organism	Haploid genome size (n)	Expected number of fragments		
		4-Cutter (n/256)	6-Cutter (n/4096)	8-Cutter (n/65 536)
Escherichia coli	4.7×10^6 bp	18 359	1147	72
Saccharomyces cerevisiae	1.35×10^7 bp	52 734	3296	206
Drosophila melanogaster	1.8×10^8 bp	703 125	43 945	2746
Homo sapiens	2.8×10^9 bp	1093 750	683 593	42 274

Table 3.1 Expected number of DNA fragments produced from different genomes by restriction endonucleases with tetra-, hexa- and octanucleotide recognition sequences. The expected number assumes that the DNA has a 50% G+C content and there is a totally random distribution of the four bases along any one strand of DNA.

the fragments range in size from 4.5 kb to 3.5 Mb (Fan *et al.* 1989). If more uniform-sized fragments are required it is usual to *partially* digest the target DNA with an enzyme with a 4-base recognition sequence. The partial digest then can be fractionated (see next section) to separate out fragments of the desired size. Since the DNA is randomly fragmented there will be no exclusion of any sequence. Furthermore, clones will overlap one another (Fig. 3.1) and this is particularly important when trying to order different clones into a map (see Chapter 4).

Some introns encode endonucleases that are site-specific and have 18–30 bp recognition sequences (Dujon *et al.* 1989). These endonucleases can be used to produce a very limited number of fragments, some or all of which are produced by cleavage within related genes. For example, the intron-encoded endonuclease I-*Cen*I from the chloroplast large rRNA gene of *Chlamydomonas eugametos* cuts the chromosomes of *E. coli* and *Salmonella* seven times, once within each of the seven *rrn* operons (Liu *et al.* 1993).

New sites for rare-cutting endonucleases can be introduced into genomes to facilitate gene mapping. For example, in bacteria a number of temperate phages and transposons contain single *Not*I sites and their insertion into the chromosome will create one or more new sites depending on the number of integration events (Smith *et al.* 1987; Le Bourgeois *et al.* 1992). Similarly, an integrative vector was used to generate sites in yeast for the intron-encoded endonuclease I-*Sce*I (Thierry & Dujon 1992). A variation of this technique can be used to reduce the number of cleavage sites. A novel *E. coli lac* operator containing a site for *Hae*II (Pu GCGC Py) was introduced into the genomes of *E. coli* and yeast. Addition of the *lac* repressor protein protects this site from methylation by the M.*Hha* methyltransferase which recognizes the sequence GCGC found in all *Hae*II recognition sites. Consequently, a single site remains that is susceptible to cleavage by *Hae*II, the so-called *Achilles heel* site (Koob & Szbalski 1990). The RecA-assisted restriction endonuclease (RARE) technique for reducing the number of cleavage

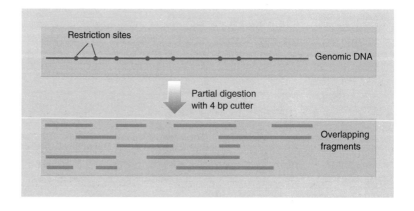

Fig. 3.1 The generation of overlapping fragments of genomic DNA following partial digestion with a restriction enyzme with a tetranucleotide recognition sequence.

sites involves the RecA-mediated formation of a triplex structure between an oligonucleotide and a chosen locus on the chromosome containing the site of a given restriction enzyme (Koob & Szbalski 1990; Ferrin & Camerini-Otero 1991). After methylation of the genome by the corresponding methyltransferase and removal of the RecA protein, only the protected site is cleaved by the endonuclease. Although these methods for reducing the number of restriction sites potentially are very powerful, they are technically difficult and are not widely used.

Separating large fragments of DNA

As a DNA molecule is digested with a restriction endonuclease it is cut into smaller and smaller pieces. As such, the number of smaller molecules will always exceed the number of larger molecules. Small fragments ligate more efficiently and clones with small inserts transform with a higher efficiency. If there is no size selection, then in a random cloning exercise the recombinants will have a preponderance of small inserts. Some vectors have an automatic size selection, e.g. the λ phage (5–25 kb) and cosmid (35–45 kb) vectors, but most do not. Thus it is common practice after DNA digestion to isolate DNA fragments of the desired size.

One method of achieving size separation of DNA fragments is by centrifugation through sucrose density gradients. There are several ways of making a sucrose density gradient but the commonest method is the use of a gradient-maker (Fig. 3.2). This consists of two reservoirs joined by a capillary tube which is fitted with a valve. In addition, one of the reservoirs has an outlet tube and contains a stirring device. If a 5–20% sucrose gradient is desired, 20% sucrose is put into the mixing reservoir and 5% sucrose in the other. Stirring is begun and the exit line opened. As 20% sucrose leaves the mixing reservoir the valve separating the two reservoirs is opened to allow 5% sucrose to flow in and gradually dilute the 20% sucrose. A small volume of macromolecule-containing solution is then gently layered on top of the gradient. Upon application of a centrifugal field, the macro-molecules sediment through the solution at a rate dependent on their size and shape. The rate of sedimentation is usually expressed as the sedimentation coefficient, S :

$$S = \frac{dx/dt}{\omega^2 x}$$

where x is the distance from the centre of rotation, ω the angular velocity in radians per second, and t the time in seconds. Nucleic acids have sedimentation coefficients in the range between 1 and 100×10^{-13} s. A sedimentation coefficient of 1×10^{-13} s is called a Svedberg and is abbreviated S. Thus a sedimentation coefficient of 42×10^{-13} s would be denoted 42 S.

Figure 3.2 Preparation and use of sucrose density gradients. (a) Apparatus used for the preparation of sucrose density gradients (see text for details). (b) Separation of two components in a sucrose density gradient. At time t_0 the mixture of two components is layered on top of the gradient. After centrifugation for time t_1 the two components have separated and by time t_2 the faster of the components has pelleted. Note that if centrifugation is continued indefinitely both components will be pelleted and separation will not be achieved.

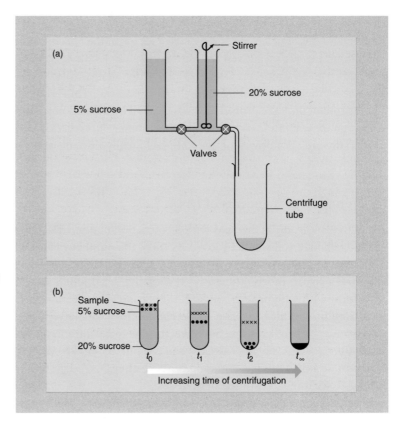

There is no theoretical reason why macromolecules cannot be separated by sedimentation through a column of water instead of sucrose. Practically, however, good separations are hard to achieve in this way because the slightest disturbance of the centrifuge tube will cause remixing; the presence of a gradient of sucrose minimizes such disturbances. In addition, the increasing concentration of sucrose counteracts the increasing centrifugal force imposed on nucleic acids as they move further from the centre of rotation; in this way the rate of sedimentation is kept constant.

The alternative method of separating large DNA molecules is by electrophoresis in agarose. An agarose gel is a complex network of polymeric molecules whose average pore size is dependent on the buffer composition and the type and composition of agarose used. Upon electrophoresis, DNA molecules display elastic behaviour by stretching in the direction of the applied field and then contracting into balls. The larger the pore size of the gel, the greater the ball of DNA that can pass through and hence the larger the molecules that can be separated i.e. the gel acts as a sieve. Once the globular volume of the DNA molecule exceeds the pore size, the DNA molecule can only pass through by reptation (i.e. a snake-like movement). This leads to size-independent mobility which occurs with conventional agarose gel electrophoresis above 20–30 kb. To achieve reproducible separations of DNA fragments larger than this it is necessary to use pulsed electrical fields.

Pulsed field gel electrophoresis (PFGE) was developed by Schwartz and Cantor (1984). The technique employs alternately pulsed perpendicularly oriented electrical fields (Fig. 3.3). The duration of the applied electrical pulses is varied from 1 s to 90 s permitting separation of DNAs with sizes from 30 kb to 2 Mb. The mechanism by which this occurs is believed to be as follows. When a large DNA molecule enters a gel in response to an electrical field it must elongate parallel to the field. This field then is cut off and a new field applied at right angles to the long axis of the DNA. The molecule is unable to move until it re-orientates in a new direction and the time required for this will depend on molecular weight. Repeating the cycle results in each DNA molecule having a characteristic net mobility along the diagonal of the gel. The use of non-uniform electrical fields is critical in achieving high resolution. The reason for this is that the leading edge of a DNA band is exposed to a weaker electrical field than the trailing edge and thus the band is subject to constant compression. The usefulness of this method has been demonstrated by the separation of the different fragments of DNA produced by *Not*I digestion of *E. coli* and *S. pombe* DNA (Smith *et al.* 1987; Fan *et al.* 1989) and the separation of intact *S. cerevisiae* chromosomes (Schwartz & Cantor 1984).

A major disadvantage of PFGE, as originally described, is that the DNA samples do not run in straight lines making analysis by Southern

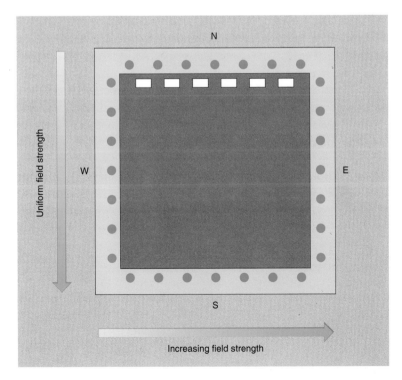

Figure 3.3 Instrumentation for pulse field gradient electrophoresis. The platinum electrodes are shown as green dots and the gel slots are positioned along one edge of the gel matrix which is shown in grey. The letters N, S, E and W refer to the electrical field orientation. (Adapted from Schwartz & Cantor 1984.)

blotting, etc. difficult. This problem has been overcome by the development of improved methods for alternating electrical fields including orthogonal-field-alternation gel electrophoresis (OFAGE), field-inversion gel electrophoresis (FIGE), transversely alternating-field electrophoresis (TAFE) and contour-clamped homogeneous electrical field (CHEF) (Cantor *et al.* 1988). CHEF, which was first described by Chu *et al.* (1986), is the most commonly used method for genomes of all types. It uses a hexagonal array of fixed electrodes (Fig. 3.4) and this creates a homogeneous electrical field resulting in enhanced resolution of DNA fragments. The DNA tracks run in straight lines and are comparable across the gel, thus making calibration with markers and the interpretation of Southern blots simpler than with the original PFGE.

Isolation of chromosomes

The task of lining up the different clones that make up a gene library can be simplified if the individual chromosomes are separated prior to digestion. PFGE permits the separation of yeast chromosomes (Fan *et al.* 1989), the largest of which is 3–5 Mb. However, the chromosomes of higher eukaryotes are much larger than this. For example, *Drosophila* chromosome 2 is 67 Mb in length and human chromosomes vary in size from 50 Mb (chromosome 21) to 263 Mb (chromosome 1). Electrophoresis cannot separate molecules this large. Instead, fluorescence activated cell sorting (FACS), also known as flow karyotyping, has to be used (Davies *et al.* 1981).

In FACS, chromosome preparations are stained with a DNA

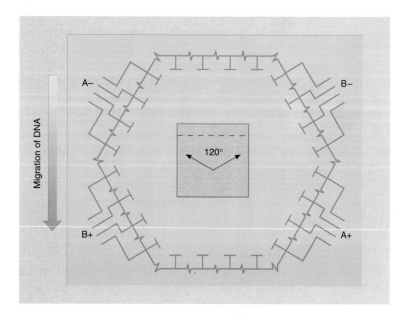

Figure 3.4 Schematic representation of CHEF (contour-clamped homogeneous electrical field) pulsed field gel electrophoresis.

binding dye which can fluoresce in a laser beam. The amount of fluorescence exhibited by a given chromosome is proportional to the amount of dye bound. This in turn is proportional to the amount of the DNA and hence the size of the chromosome. Chromosomes can therefore be fractionated by size in a FACS machine (Fig. 3.5). A stream of droplets containing single stained chromosomes is passed through a laser beam at the rate of 2000 chromosomes per second and the fluorescence from each is monitored by a photomultiplier. When the fluorescence intensity indicates that the chromosome illuminated by the laser is the one desired, the charging collar puts an electrical charge on the droplet. When droplets containing the desired chromosome pass between charged deflection plates, they are deflected into a collection vessel. Uncharged droplets lacking the desired chromosome pass into a waste collection vessel. In a variation of the above technique, two fluorescent dyes are used: one binds preferentially to AT-rich DNA and the other binds preferentially to GC-rich DNA. The stained chromosomes pass through a point on which a pair of laser beams are focused, one beam to excite the fluorescence of each dye. Each chromosome type has characteristic numbers of AT and GC base pairs so chromosomes can be identified by a combination of the total fluorescence and the ratio of the intensities of the fluorescence emissions from the two dyes.

The kind of separation that can be achieved with human chromosomes is shown in Fig. 3.6. Note that some of the chromosomes cannot be separated in the FACS because they are of similar size and AT/GC ratio. However, separation can be achieved if *somatic cell hybrids* are used as the source of DNA (D'Eustachio & Ruddle 1983). These hybrids contain a full complement of chromosomes of one species but only one or a limited number of chromosomes from the second. They are formed by fusing the cells of the two different species and applying conditions that select against the two donor cells. For example, one of the parents may be sensitive to a particular drug and the other might be a mutant requiring special conditions for growth, e.g. thymidine kinase negative cells do not grow in HAT medium. In the presence of the drug and HAT medium, only hybrid cells containing a functional thymidine kinase gene can grow. If hybrid cells are grown under non-selective conditions after the initial selection, chromosomes from one of the parents tend to be lost more or less at random. In the case of human/rodent fusions, which are the most common, the human chromosomes are preferentially lost. Eventually only one or a few chromosomes from one parent remain. In this way rodent cell lines containing one or two human chromosomes have been constructed and these can be used to isolate the individual human chromosomes by FACS. Lee *et al.* (1994) have shown that chromosome sorting from hybrid cells can be simplified if rodent cells are replaced with cells from the Indian Muntjac deer. Cells of the Muntjac deer have only a very small number of giant

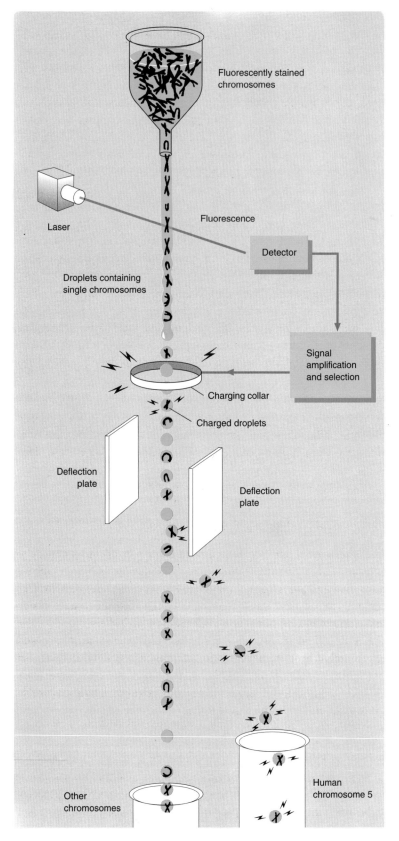

Fluorescently stained chromosomes

Laser

Fluorescence

Detector

Droplets containing single chromosomes

Signal amplification and selection

Charging collar

Charged droplets

Deflection plate

Deflection plate

Other chromosomes

Human chromosome 5

Figure 3.5 Schematic representation of single chromosome separation b FACS. See text for details. (Redrawn with permissior from Los Alomos Science No. 20 courtesy of University Science Books.

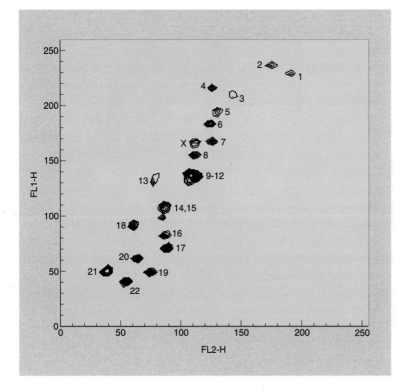

Figure 3.6 Separation of different human chromosomes by FACS using two fluorescent dyes, Hoechst 33258 (FL1-H) and chromomycin A3 (FL2-H). (Kindly supplied by Dr D. Davies, Imperial Cancer Research Fund.)

chromosomes: two autosomes plus X and Y. Thus donor chromosomes in hybrid cells are easily separated from host chromosomes. Not only were highly purified human chromosomes obtained by flow cytometry but they were obtained at over 90% purity by rate zonal centrifugation in sucrose gradients.

Chromosome microdissection

Particular subchromosomal regions, such as a chromosomal band (see Fig. 2.20), can be obtained by microdissection of metaphase chromosomes. A micromanipulator with very fine needles is used to cut out the desired band from individual chromosomes. When sufficient material has been collected the DNA is cloned. The technique originally was developed for use with *Drosophila* and mouse DNA and at least 100 chromosomes were needed for dissection. Ludecke *et al.* (1989) have simplified this technique. Only a few chromosomes are dissected and after cloning into a vector this DNA is amplified by means of PCR.

Vectors for cloning DNA

Once the genome has been fragmented it is essential to propagate each fragment to enable it to be mapped and ultimately sequenced. Although many different types of vector have been developed (see

47

Old & Primrose 1994, for review) only a few are of use for large-scale genome sequencing projects. The early work on construction of gene libraries made use of bacteriophage λ vectors. With such vectors the largest size of DNA that can be accommodated is about 25 kb. If *in vitro* packaging is used then a large number of independent recombinants can be selected.

In place of the phage λ vectors, cosmid vectors may be chosen. These also have the high efficiency afforded by packaging *in vitro* and have an even higher capacity than any phage λ vector. However, there are two drawbacks in practice. First, most workers find that screening libraries of phage λ recombinants by plaque hybridization gives cleaner results than screening libraries of bacteria containing cosmid recombinants by colony hybridization. Plaques usually give less of a background hybridization than do colonies. Second, it may be desired to retain and store an amplified genomic library. With phage, the initial recombinant DNA population is packaged and plated out. It can be screened at this stage. Alternatively, the plates containing the recombinant plaques can be washed to give an *amplified* library of recombinant phage. The amplified library can then be stored almost indefinitely; phage have a long shelf-life. The amplification is so great that samples of this amplified library could be plated out and screened with different probes on hundreds of occasions. With bacterial colonies containing cosmids it is also possible to store an amplified library (Hanahan & Meselson 1980), but bacterial populations cannot be stored as readily as phage populations. There is often an unacceptable loss of viability when the bacteria are stored.

A word of caution is necessary when considering the use of any amplified library. There is the possibility of *distortion*. Not all recombinants in a population will propagate equally well, e.g. variations in target DNA size or sequence may affect replication of a recombinant phage, plasmid or cosmid. Therefore, when a library is put through an amplification step particular recombinants may be increased in frequency, decreased in frequency, or lost altogether. Development of modern vectors and cloning strategies has simplified library construction to the point where many workers now prefer to create a new library for each screening, rather than risk using a previously amplified one.

Genomic DNA libraries in phage λ vectors are expected to contain most of the sequences of the genome from which they have been derived. However, deletions can occur, particularly with DNA containing repeated sequences. As noted in Chapter 2, repeated sequences are widespread in the DNA from higher eukaryotes. The deletion of these repeated sequences is not prevented by the use of recombination-deficient strains.

Yeast artificial chromosomes (YACs)

The upper size limit of 35–45 kb for cloning in a cosmid means that it would take 4500 clones to cover the *D. melanogaster* genome and 70 000 clones to cover the human genome. What is required is a vector that permits a larger insert size and YACs make this possible. It will be recalled from Chapter 2 that the minimum structural elements for a linear chromosome are an origin of replication (*ars*), telomeres and a centromere. Murray and Szostak (1983) combined an *ars* and a centromere from yeast with telomeres from *Tetrahymena* to generate a linear molecule that behaved as a chromosome in yeast. When the requirements for normal replication and segregation were studied it was found that the length of the YAC was important. When the YAC was less than 20 kb in size, centromere function was impaired. However, much larger YACs segregated normally. Burke *et al.* (1987) made use of this fact in developing a vector (Fig. 3.7) for cloning large DNA molecules. They showed that YACs could be used to generate whole libraries from the genomes of higher organisms with insert sizes at least tenfold larger than those that can be accommodated by bacteriophage λ and cosmid vectors.

A major drawback with YACs is that although they are capable of replicating large fragments of DNA, manipulating such large DNA fragments in the liquid phase prior to transformation, and keeping them intact, is very difficult. Thus many of the early YAC libraries had average insert sizes of only 50–100 kb. By removing small DNA fragments by PFGE fractionation prior to cloning, Anand *et al.* (1990) were able to increase the average insert size to 350 kb. By including polyamines to prevent DNA degradation, Larin *et al.* (1991) were able to construct YAC libraries from mouse and human DNA with average insert sizes of 700 and 620 kb, respectively. More recently, Bellanné-Chantelot *et al.* (1992) constructed a human library with an average insert size of 810 kb and with some inserts as large as 1800 kb.

There are a number of operational problems associated with the use of YACs (Kouprina *et al.* 1994; Monaco & Larin 1994). The first of these is that it is estimated that 10–60% of clones in existing libraries represent chimaeric DNA sequences: that is, sequences from different regions of the genome cloned into a single YAC. Chimaeras may arise by co-ligation of DNA inserts *in vitro* prior to yeast transformation, or by recombination between two DNA molecules that were introduced into the same yeast cell. It is possible to detect chimaeras by *in situ* hybridization of the YAC to metaphase chromosomes: hybridization to two or more chromosomes or to geographically disparate regions of the same chromosome is indicative of a chimaera.

A second problem with YACs is that many clones are unstable and tend to delete internal regions from their inserts. Using a model

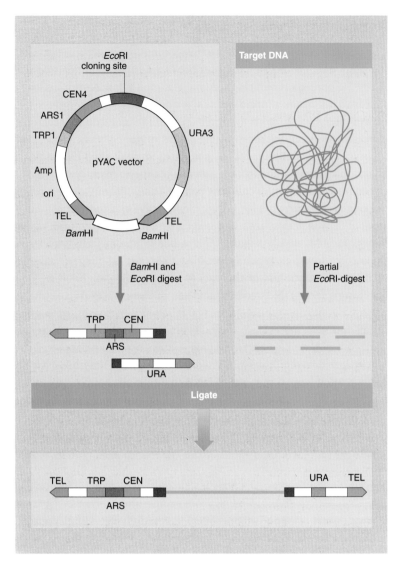

Figure 3.7 Construction of a YAC containing large pieces of cloned DNA. Key regions of the pYAC vector are as follows: TEL, yeast telomeres; ARS1, autonomously replicating sequence; CEN4, centromere from chromosome 4; URA3 and TRP1, yeast marker genes; Amp, ampicillin-resistance determinant of pBR322; ori, origin of replication of pBR322.

system, Kouprina *et al.* (1994) were able to show that deletions can be generated both during the transformation process and during mitotic growth of transformants and that the size of the deletions varied from 20 to 260 kb. The frequency of deletion formation could be reduced, but not eliminated, by use of a recombination deficient strain (Ling *et al.* 1993). Heale *et al.* (1994) have shown that chimaera formation results from the yeast's mitotic recombination system which is stimulated by the spheroplasting step of the standard YAC transformation system. Transformation of intact yeast cells is much less recombinagenic. An additional limitation on the use of YACs is the high rate of loss of some YACs during mitotic growth.

The third major problem with YAC clones is that the 15 Mb yeast host chromosome background cannot be separated from the YACs by simple methods. Nor is the yield of DNA very high. Unlike plasmid vectors in bacteria, YACs have a structure very similar to natural yeast chromosomes. Thus purifying YAC from the yeast chromosomes usually requires separation by PFGE. Alternatively, the entire yeast genome is subcloned in bacteriophage or cosmid vectors followed by identification of those clones carrying the original YAC insert.

BACs and PACs as alternatives to YACs

Phage P1 is a temperate bacteriophage which has been extensively used for genetic analysis of *E. coli* because it can mediate generalized transduction. Sternberg and co-workers have developed a P1 vector system which has a capacity for DNA fragments as large as 100 kb (Sternberg 1990; Pierce *et al.* 1992). Thus the capacity is about twice that of cosmid clones but less than that of YAC clones. The P1 vector contains a packaging site (pac) which is necessary for *in vitro* packaging of recombinant molecules into phage particles. Vectors contain two *lox*P sites. These are the sites recognized by the phage recombinase, the product of the phage *cre* gene, and which lead to circularization of the packaged DNA after it has been injected into an *E. coli* host expressing the recombinase (Fig. 3.8). Clones are maintained in *E. coli* as low copy number plasmids by selection for a vector kanamycin-resistance marker. A high copy number can be induced by exploitation of the P1 lytic replicon (Sternberg 1990). This P1 system has been used to construct genomic libraries of mouse, human, fission yeast and *Drosophila* DNA (Hoheisel *et al.* 1993; Hartl *et al.* 1994). Estimates of chimaerism in such libraries are well below the estimates for YAC libraries. For example, Hartl *et al.* (1994) found chimaeric clones in less than 2% of more than 3000 *Drosophila* P1 clones and Sternberg (1994) reported less than 5% in more than 1000 human and mouse P1 clones.

Shizuya *et al.* (1992) have developed a bacterial cloning system for mapping and analysis of complex genomes. This BAC system (*bacterial artificial chromosome*) is based on the single-copy sex factor F of *E. coli*. This vector (Fig. 3.9) includes the λ *cos*N and P1 *lox*P sites, two cloning sites (*Hind*III and *Bam*HI) and several G+C restriction enzyme sites (e.g. *Sfi*I, *Not*I, etc.) for potential excision of the inserts. The cloning site also is flanked by T7 and SP6 promoters for generating RNA probes. This BAC can be transformed into *E. coli* very efficiently thus avoiding the packaging extracts that are required with the P1 system. The BAC is capable of maintaining human and plant genomic fragments of greater than 300 kb for over 100 generations with a high degree of stability (Woo *et al.* 1994) and has been used to construct a rice nuclear genome library with an average insert size of 125 kb (Wang *et al.* 1995a). There are, however, two disadvantages to the use of BACs. First, there currently is no method of positively selecting clones with foreign DNA inserts. Secondly, because BACs cannot be amplified it is difficult to isolate large amounts of DNA.

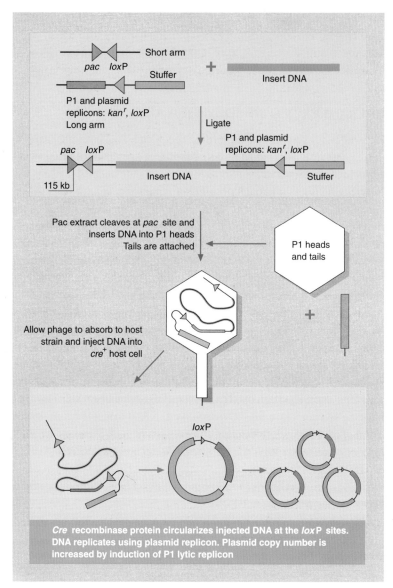

Figure 3.8 The phage P1 vector system. The P1 vector Ad10 (Sternberg 1990) is digested to generate short and long vector arms. These are dephosphorylated to prevent self-ligation. Size-selected insert DNA (85–100 kb) is ligated with vector arms, ready for a two-stage processing by packaging extracts. First, the recombinant DNA is cleaved at the *pac* site by pacase in the packaging extract. Then the pacase works in concert with head/tail extract to insert DNA into phage heads, *pac* site first, cleaving off a headful of DNA at 115 kb. Heads and tails then unite. The resulting phage particle can inject recombinant DNA into host *E. coli*. The host is *cre*+. The *cre* recombinase acts on *lox*P sites to produce a circular plasmid. The plasmid is maintained at low copy number, but can be amplified by inducing the P1 lytic operon.

More recently, Ioannou *et al.* (1994) have developed a P1-derived artificial chromosome, or PAC, by combining features of both the P1 and F-factor systems. The PAC vector is able to handle inserts in the 100–300 kb range and as yet no chimaeras or clone instability have been detected.

Human artificial episomal chromosomes (HAECs)

One problem with YACs, BACs and PACs is that they are not maintained as plasmids or minichromosomes outside their 'normal' host, i.e. yeast or *E. coli*. Rather, in foreign cells the vector DNA plus insert can be maintained only by integrating into the host

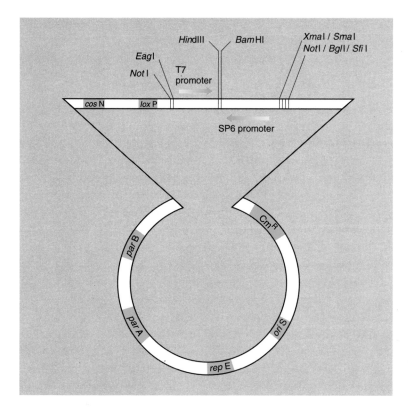

Figure 3.9 Structure of a BAC vector derived from a mini-F plasmid. The *ori*S and *rep*E genes mediate the unidirectional replication of the F factor while *par*A and *par*B maintain the copy number at a level of one or two per genome. Cm^R is a chloramphenicol-resistance marker. *cos*N and *lox*P are the cleavage sites for λ terminase and P1 *cre* protein, respectively. *Hind*III and *Bam*HI are unique cleavage sites for inserting foreign DNA. (Adapted from Shizuya *et al.* 1992.)

chromosome. This is not ideal for studying the function of the cloned sequence, particularly if it is a functional unit many hundreds of kilobases in length. Furthermore, integration can result in damage to chromosomal genes and the integrated DNA may be rearranged, deleted or disrupted. To try and overcome this problem a number of groups have attempted, unsuccessfully (Brown 1992), to construct mammalian artificial chromosomes, or big MACs! Now Sun *et al.* (1994) have developed a HAEC system based on the replication origin of the herpes Epstein–Barr virus. HAECs are stably maintained in human cells as circular minichromosomes and so far have been used to carry inserts as large as 330 kb.

Choice of vector

The maximum size of insert that the different vectors will accommodate is shown in Table 3.2, and Table 3.3 shows the size of different genomes relative to the size of insert with the various vectors. However, as noted above, the size of insert is not the only feature of importance. The absence of chimaeras and deletions is even more important. Cosmids are better than YACs in this respect and they have an additional advantage: it is easy to isolate large amounts of cosmid DNA. Potential advantages of the BAC and PAC systems over YACs include lower levels of chimaerism, ease of library generation and ease of manipulation and isolation of insert DNA (Woo *et al.* 1994).

Avoiding the need to isolate individual chromosomes

The construction and identification of YAC clones from specific chromosomes serves as an important step in the assembly of overlapping clone libraries of these chromosomes. Earlier (p. 45) it was described how individual chromosomes could be isolated by flow cytometry prior to cloning. However, the small amount of material generated by flow-sorting makes library construction with very large inserts extremely difficult. This leads to an increase in the number of clones required to achieve a given coverage, which partly offsets the advantage of having a chromosome-specific library.

An alternative approach is to make a YAC library from a somatic cell hybrid (see p. 45), e.g. a monosomic human–rodent hybrid. Once the library is constructed those clones that carry the human DNA are distinguished from those containing rodent DNA by hybridization with a probe specific for the former, e.g. human DNA repeat sequences (Abidi *et al.* 1990; Wada *et al.* 1990). The problem with this approach is that to isolate all 24 human chromosomes (22 autosomes plus two sex chromosomes) would require the construction of 24 individual libraries from a full panel of human monosomic hybrid cell lines. The only justification for such a daunting task is that YAC libraries derived from rodent–human cell lines *might*

Table 3.2 Maximum DNA insert possible with different cloning vectors.

Vector	Host	Insert size
λ phage	*E. coli*	5 – 25 kb
λ cosmids	*E. coli*	35 – 45 kb
P1 phage	*E. coli*	70 – 100 kb
PACs	*E. coli*	100 – 300 kb
BACs	*E. coli*	≤ 300 kb
HAECs	Mammalian cells	≤ 330 kb
YACs	*S. cerevisiae*	200 – 2000 kb

Table 3.3 Size of different genomes relative to the average DNA insert that is obtained with different cloning vectors.

Organism	Haploid genome size	Size of genome relative to size of cloned DNA (n)			
		Cosmid (40 kb)	P1 (85 kb)	BAC/PAC (250 kb)	YAC (1000 kb)
E. coli	4.7×10^5 bp	118	55	19	5
S. cerevisiae	1.35×10^7 bp	338	159	54	14
D. melanogaster	1.8×10^8 bp	4 500	2 118	720	180
H. sapiens	2.8×10^9 bp	70 000	32 941	11 200	2800

contain a lower frequency of chimaeric YACs than those derived from a single species.

A less labour-intensive approach than the one described above is to generate a complete genomic library and then divide it into chromosomal subsets. Division is achieved by screening all the clones with chromosome-specific DNA probes. The feasibility of this approach has been proven with human chromosome 21. Ross *et al.* (1992) developed their chromosome-21-specific probes from a series of phage libraries constructed from flow-sorted chromosomes. Inserts were excised from vector sequences and recloned using a plasmid vector. The plasmid clones from each chromosome then were pooled together. The purpose of this recloning procedure was to increase the ratio of insert to vector sequences. Chumakov *et al.* (1992a) developed their probes from somatic cell hybrid lines containing individual human chromosomes. Human chromosome DNA lying between *Alu* repeats was amplified by PCR using as a primer a consensus *Alu* repeat sequence. In this way a family of chromosome-21-specific sequences was obtained free of rodent DNA. Yet a third set of chromosome-21-specific probes was developed by Chumakov *et al.* (1992b) from the known sequence of chromosome 21 genes deposited in nucleic acid databases.

4 Assembling a physical map of the genome

Introduction

After a genome has been fragmented and the fragments cloned to generate a genomic library, it is necessary to assemble the cloned fragments in the same linear order as found in the chromosomes from which they were derived. Positioning cloned DNA fragments is analogous to completing one edge of a jigsaw puzzle but, rather than looking for interlocking pieces, detectable overlaps between clones are looked for, i.e. clones with a unique stretch of DNA in common. Because the number of fragments is so large a rapid method for detecting overlaps between pairs of clones is needed. If each clone could be sequenced, overlaps could be identified unambiguously, provided the overlapping region is not a sequence that repeats itself elsewhere in the genome. However, the current state of sequencing technology is such that this approach is totally impractical. For example, using a cosmid library of the human genome in which the average insert size is 40 kb, approximately 10 000 sequencing gels would be required.

One method of linking cloned fragments is *chromosome walking* (Bender *et al.* 1983) which was originally developed for the isolation of gene sequences whose function is unknown but whose genetic location is known. The principle of this method is shown in Fig. 4.1. For the purposes of map generation a single cloned fragment is selected. This is used as a probe to detect other clones in the library with which it will hybridize and which represent clones overlapping with it. The overlap can be to the right or the left. This single walking step is repeated many times and can occur in both directions along the chromosome. A potential problem with chromosome walking is created by the existence of repeated sequences. If the clone used as the probe contains a sequence repeated elsewhere in the genome it will hybridize to non-contiguous fragments. For this reason the probe used for stepping from one genomic clone to the next must be a unique sequence clone, or a subclone which has been shown to contain only a unique sequence. Clearly, if chromosome walking is to be employed it makes sense to use very large fragments of DNA as this minimizes the number of steps.

56

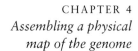

CHAPTER 4
*Assembling a physical
map of the genome*

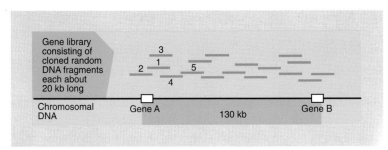

Figure 4.1 Chromosome walking. It is desired to clone DNA sequences of gene B, which has been identified genetically but for which no probe is available. Sequences of a nearby gene A are available in fragment 1. Alternatively, a sequence close to gene B could be identified by *in situ* hybridization to *Drosophila* polytene chromosomes. In a large, random genomic DNA library many overlapping cloned fragments are present. Clone 1 sequences can be used as a probe to identify overlapping clones 2, 3 and 4. Clone 4 can, in turn, be used as a probe to identify clone 5, and so on. It is, therefore, possible to walk along the chromosome until gene B is reached.

Although inherently attractive, chromosome walking is too laborious and time consuming to be of value for mapping of most genomes. Reference to Table 3.3. shows that with cosmid clones a minimum of 4500 individual 'walks' would be necessary with the *Drosophila* genome and 70 000 with the human genome. In practice, many more 'walks' would be necessary since a representative library would contain a much larger number of clones (see p. 38). Three general methodologies for mapping have been developed as alter-natives to chromosome walking: restriction enzyme fingerprinting, marker sequences and hybridization assays. In each case the objective is to create a landmark map of the genome under study with markers dispersed at regular intervals throughout the map. In restriction enzyme fingerprints the markers are restriction sites; with the other methods they are unique sequences detected by the PCR or by hybridization. Each of the methods has its limitations and the current trend is to integrate the maps produced by the different methods.

A physical map of the *E. coli* chromosome

The principle of restriction enzyme fingerprinting was originally developed for the nematode *Caenorhabditis elegans* (Coulson *et al.* 1986) and yeast (Olson *et al.* 1986). However, the first construction of a complete physical map was that of *E. coli*. It is a particularly elegant example and has been selected here for illustrative purposes. The starting point for the *E. coli* map (Kohara *et al.* 1987) was the construction of a genomic library in a lytic phage λ vector. Although cosmid clones contain much longer stretches of cloned DNA they frequently accumulate deletions of *E. coli* DNA because of metabolic

CHAPTER 4
*Assembling a physical
map of the genome*

imbalances that affect the growth of host cells caused by increased dosage of particular genes on the cosmid. A lytic λ vector has a smaller cloning capacity but has the advantage that the cloned fragments will be stably maintained because it kills host bacteria within a short time.

Once the genomic library had been constructed each clone was individually restriction enzyme fingerprinted. For this purpose, DNA from each clone was partially digested in separate reactions with eight different restriction enzymes. The fragments produced were separated by agarose gel electrophoresis. Each gel was subjected to Southern blotting using as a probe a radioactively-labelled 2.7 kb fragment from the λ vector. The pattern of bands detected was then read in an analogous fashion to a DNA sequencing gel (Fig. 4.2). Starting with 1056 clones, this step was completed in 4 weeks and usable data were generated for 1025 clones.

The restriction fingerprint of each clone then was compared with that of all other clones. Overlapping clones were selected on the basis of a match of at least five consecutive cleavage sites (Fig. 4.2). In this way the 1025 clones were sorted into 70 groups including seven stand-alone clones. The size of the groups ranged from 20 to 180 kb and their sum amounted to 4.4 Mb, i.e. 94% of the genome. Since the average insert size of the library was 15.5 kb, 1025 clones should theoretically represent 96% of the genome (see p. 38). This suggested that the gap between each group could not be larger than several kilobases in length. Thus it should be possible to close each gap with a single clone.

One way of closing the gaps in the map would be to isolate a fresh set of clones and fingerprint them. However, this would be labour intensive. Instead the clones at the ends of each group were used as probes to identify all the other clones with which they would hybridize. For example, clone 6E8, which had been located at the end of group 33, hybridized to clones 22E3 and 14E10. Clone 22E3 also hybridized to clone 9F3 which was located at the end of group 6. Thus clone 22E3 could bridge groups 33 and 6 and this was confirmed by analysis of its restriction fingerprint (Fig. 4.3).

The availability of a complete physical map like that of Kohara *et al.* (1987) is a very powerful tool. For example, computer analysis of the nucleotide sequence of a cloned gene allows a simple restriction map to be derived and hence located on the physical map. For *E. coli* sequence data two groups have developed appropriate programmes. One is based on restriction fragment alignment (Rudd *et al.* 1990, 1991) and the other on the comparison of the length of each restriction fragment (Médigue *et al.* 1990, 1993).

Other examples of restriction fragment mapping

The fingerprinting technique described above for mapping the *E.*

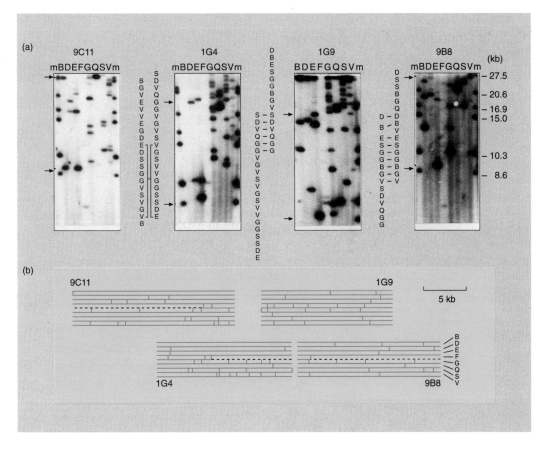

Figure 4.2 Example of restriction fragment mapping of cloned *E. coli* DNA. (a) Clones of *E. coli* DNA were partially digested with restriction enzymes *Bam*HI(B), *Hin*dIII(D), *Eco*RI(E), *Eco*RV(F), *Bgl*II(G), *Kpn*I(Q), *Pst*I(S) and *Pvu*II(V) and the fragments were separated by electrophoresis. The autoradiograms of four clones are shown after Southern blotting and hybridization with a λ vector DNA probe. In each electropherogram, m represents the lane for the size markers. The boundaries of the cloned DNA and the vector are indicated by arrows. The cleavage sites for each enzyme were 'read' and aligned. The restriction maps deduced from the autoradiograms are depicted in (b). Rightward direction in the maps corresponds to upward (1G4, 1G9 and 9B8) or downward (9C11) in the autoradiograms. Broken lines in the map indicate that the restriction enzyme cleavage sites above them were unable to be read out from the autoradiograms. The right-most *Bam*HI cleavage site found in clone 9C11 was not found in an overlapping clone, 1G4. It is likely that this site had been created in clone 9C11 at the time of library construction by ligating a particular *Sau*3A end such as GATCC into the *Bam*HI site of the vector. (Redrawn with permission from Kohara *et al.* 1987, copyright (1987) Cell Press.)

coli genome has been used, albeit with some variation, with the genomes of other microbes (see Cole & Saint Girons 1994 for review) as well as eukaryotic organisms, e.g. *Saccharomyces cerevisiae* (Olson *et al.* 1986), *Caenorhabditis elegans* (Coulson *et al.* 1986), *Drosophila melanogaster* (Sidén-Kiamos *et al.* 1990), *Arabidopsis thaliana* (Hauge *et al.* 1991), human chromosome 19 (Trask *et al.*

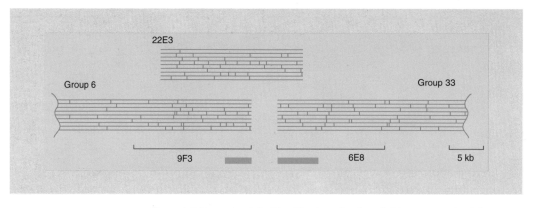

Figure 4.3 Example of the identification of a clone linking two contigs. The positions of probe DNA are indicated by solid bars. See text for details. (Redrawn with permission from Kohara *et al.* 1987, copyright (1987) Cell Press.)

1992) and the entire human genome (Bellanné-Chantelot *et al.* 1992). The original fingerprinting method devised by Coulson *et al.* (1986) is different from that of Kohara *et al.* (1987) and is shown in Fig. 4.4. Cloned DNA is digested with a restriction endonuclease with a hexanucleotide recognition sequence and which leaves staggered ends, e.g. *Hin*dIII. The ends of the fragments are labelled by end-filling with reverse transcriptase in the presence of a radioactive nucleotide triphosphate. The *Hin*dIII is destroyed by heating and the fragments cleaved again with a restriction enzyme with a tetranucleotide recognition sequence, e.g. *Sau*3A. The fragments then are separated on a high resolution gel and detected by autoradiography. In this case the fingerprint is prepared by determining the size of each band from each clone. The fingerprint obtained by this method is not an order of restriction sites. Rather, it is a series of clusters of bands based on the probability of overlap of clones.

The mapping of the *Drosophila* genome was facilitated by the existence of giant polytene chromosomes in the larval salivary glands. These polytene chromosomes exhibit a pattern of bands and interbands that is remarkably constant. Sidén-Kiamos *et al.* (1990) microdissected discrete and identifiable pieces of these polytene chromosomes and amplified them by PCR to generate probes. These probes then were used to subdivide a genomic library of cosmid clones into families of clones from the same region of the chromosome. The various members of the family then were assembled into contigs on the basis of restriction-fragment fingerprints, as described above for *C. elegans*. Once the contig maps had been prepared, Sidén-Kiamos *et al.* (1990) located them physically on the polytene chromosome by *in situ* hybridization.

In preliminary experiments in which DNA was microdissected from *D. melanogaster* chromosomes and used to probe the *D. melanogaster* master library, hybridization with dispersed repetitive DNA

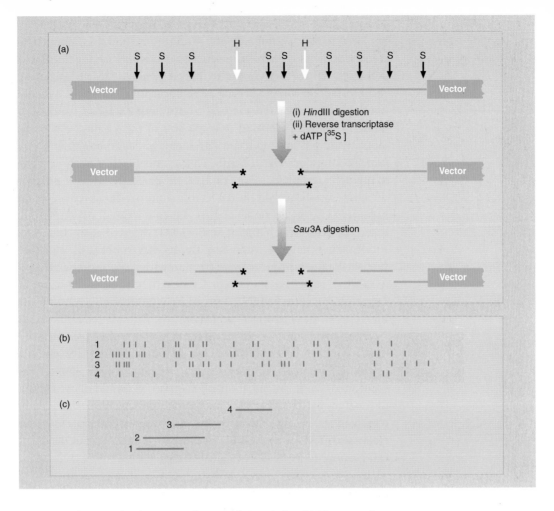

Figure 4.4 The principle of restriction-fragment fingerprinting. (a) The generation of labelled restriction fragments. See text for details. (b) Pattern generated from four different clones. Note the considerable band sharing between clones 1, 2 and 3 indicating that they are contiguous whereas clone 4 is not contiguous and has few bands in common with the other three. (c) The contig map produced from data shown in (b). (Adapted and redrawn with permission from Coulson *et al.* 1986.)

proved to be a considerable problem. This is not surprising since about 12% of the *D. melanogaster* genome is estimated to be middle repetitive DNA (see Fig. 2.6). To circumvent this problem probes were prepared from the polytene chromosomes of *Drosophila simulans* which has a much lower content of repetitive DNA and few of its repetitive sequences are found in *D. melanogaster*.

In the whole human genome approach of Bellanné-Chantelot *et al.* (1992) the contigs also were assigned to specific regions of chromosomes by *in situ* hybridization. However, the method used for assembling contigs was different. In the mapping efforts for *E. coli* and *D. melanogaster* described above, the overlap between two clones was detected by preparing a restriction-fragment fingerprint of each clone and identifying restriction-fragment lengths that are common to the two fingerprints. With this method, two clones have to overlap by at least 50% in order to declare with a high degree of certainty that the two clones do indeed overlap. Clearly, increasing the information content in each clone fingerprint would make smaller overlaps detectable. This is particularly important, as Lander and Waterman (1988) showed that the size of the smallest detectable clone overlap is an important parameter in determining the rate at which contigs increase in length. This in turn affects the rate at which mapping is completed. For example, the calculated rate of progress increases significantly if the detectable clone overlap is reduced from 50% to 25% of the clone lengths. To enable small overlaps to be detected, Bellanné-Chantelot *et al.* (1992) digested each clone with *Pvu*II, separated the fragments by gel electrophoresis and then prepared a fingerprint by hybridization with a probe prepared from a LINE-1 repeated sequence. Clones that showed little or no hybridization with the LINE-1 probe were fingerprinted with a probe based on an Alu consensus sequence.

There are a number of disadvantages with the fingerprinting approach. First, it is labour intensive and involves extensive handling of clones. Although automation helps, mapping of large genomes still is not an easy task. Second, the method quickly generates a large number of small contigs but it becomes increasingly difficult to extend and join these contigs. Finally, the method works well with phage clones but results with YACs are poor and susceptible to chimaera problems. YACs can be mapped by means of sequence tagged sites.

Sequence tagged sites

The concept of sequence tagged sites (STSs) was developed by Olson *et al.* (1989) in an attempt to systematize landmarking of the human genome. Basically, an STS is a short region of DNA about 200–300 bases long whose exact sequence is found nowhere else in the genome. Two or more clones containing the same STS must overlap and the overlap must include the STS.

Any clone that can be sequenced may be used as an STS provided it contains a unique sequence. A better method to develop STS markers is to create a chromosome-specific library in phage M13. Random M13 clones are selected and 200–400 bases sequenced. The sequence data generated are compared with all known repeated sequences to help identify regions likely to be unique. Two PCR primer sequences are selected from the unique regions which are separated by 100–300 base pairs and whose melting temperatures are similar (Fig. 4.5). Once identified, the primers are synthesized and used to PCR amplify genomic DNA from the target organism and the amplification products analysed by agarose gel electrophoresis. A functional STS marker will amplify a single target region of the genome

CHAPTER 4
*Assembling a physical
map of the genome*

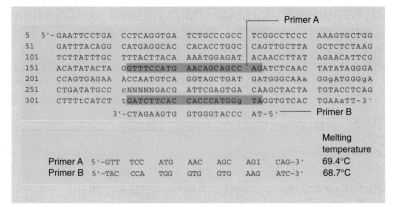
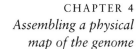

Figure 4.5 Example of an STS. The STS developed from the sequence shown above is 171 bases long. It starts at base 162 and runs through base 332. Primer A is 21 bases long and lies on the sequenced strand. Primer B is also 21 bases long and is complementary to the shaded sequence towards the 3' end of the sequenced strand. Note that the melting temperatures of the two primers are almost equal. (Reproduced with permission from Dogget 1992 courtesy of University Science Books.)

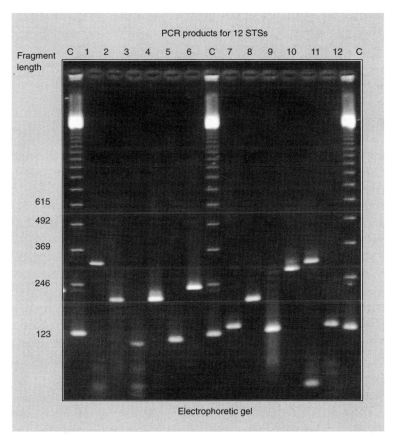

Figure 4.6 Confirmation that an STS is a unique sequence on the genome. Note that the 12 STSs from chromosome 16 shown above appear as single bands after amplification and hybridization to a chromosome 16 genomic library. (Reproduced with permission from Dogget 1992 courtesy of University Science Books.)

and produce a single band on an electrophoretic gel at a position corresponding to the size of the target region (Fig. 4.6). Alternatively, an STS marker can be used as a hybridization probe.

Operationally, an STS is specified by the sequence of the two primers that make its production possible. Thus it is defined once and for all regardless of whether the region under study is cloned in a phage, a cosmid, a BAC, PAC or YAC. Moreover, the STS will remain valid if the corresponding area of the genome is re-cloned sometime in the future in a new and, as yet, unknown vector. The STS is fully portable once the two sequences of the primer are known and these can be obtained from databanks (see Chapter 5).

The principle of the use of STSs to generate physical maps has been confirmed by a number of workers. Thirty YAC clones from the cystic fibrosis region of human chromosome 7 were assembled into a single contig (Green & Olson 1990) that spans more than 1.5 Mb. At the same time, individual YACs as large as 790 kb and containing the entire cystic fibrosis gene were constructed *in vivo* by meiotic recombination in yeast of overlapping YACs. Foote *et al.* (1992) were able to assemble 196 recombinant clones into a single over-lapping array which included over 98% of the euchromatic portion of the human Y chromosome. Similarly, Chumakov *et al.* (1992b) were able to generate an STS map for human chromosome 21q which was consistent with physical and genetic mapping data.

Radiation hybrid mapping

This method, which has been used only for mapping the human genome, makes use of somatic cell hybrids (see p. 45). A high dose of X-rays is used to break the human chromosome of interest into fragments and these fragments are recovered in rodent cells. The rodent–human hybrid clones are isolated and examined for the presence or absence of specific human DNA markers. The further apart two markers are on the chromosome, the more likely a given dose of X-rays will break the chromosome between them, placing the markers on two separate chromosomal fragments. By estimating the frequency of breakage, and thus the distance, between markers, it is possible to determine their order in a manner analogous to conventional meiotic mapping.

Radiation hybrid mapping was first developed by Goss and Harris (1975). In their experiments, human peripheral blood lymphocytes were irradiated and then fused to HPRT-deficient hamster cells. Growth in HAT medium (see p. 45) resulted in the isolation of a set of clones, each carrying a different X chromosome fragment that included the selected HPRT marker. Goss and Harris (1977) were able to establish the order of three markers on the long arm of the X chromosome and to demonstrate retention of non-selected chromosome fragments. They also derived mathematical approaches for construct-

CHAPTER 4
*Assembling a physical
map of the genome*

ing genetic maps on the basis of co-retention frequencies. However, the power of the technology could not be exploited at the time because insufficient genetic markers were available.

Renewed interest in radiation hybrid mapping was prompted by the work of Cox *et al.* (1990) who modified the original approach by using as a donor cell a rodent–human somatic cell hybrid that contained a single copy of human chromosome 21 and very little other human DNA. This cell line was exposed to 8000 rad of X-rays which resulted in an average of five human chromosome-21 pieces per cell. Because broken chromosomal ends are rapidly healed after X-irradiation, the human chromosomal fragments are usually present as translocations or insertions into hamster chromosomes. However, some cells contain a fragment consisting entirely of human chromosomal material with a human centromere. Since a dose of 8000 rad of X-rays results in cell death, the irradiated donor cells were fused with HPRT-deficient hamster recipient cells and hybrids selected as before on HAT medium. Non-selective retention of human chromosomal fragments seems to be a general phenomenon under these fusion conditions. In total, 103 independent somatic cell hybrid clones were isolated and assayed by Southern blotting for the retention of 14 DNA markers. Analysis of the results enabled the 14 markers to be mapped to a 20 Mb region of chromosome 21 and this map order was confirmed by PFGE analysis.

James *et al.* (1994) extended the work of Cox *et al.* (1990) by generating a high resolution radiation hybrid map of human chromosome 11 using 506 STSs scored on a panel of 86 radiation hybrids. A subset of 260 STSs was used to form a map with a resolution of 1 Mb between adjacent positions and which was ordered with odds of 1000:1.

As used by Cox *et al.* (1990) and James *et al.* (1994), irradiation and fusion gene transfer is a cumbersome method for creating maps of entire genomes. Barrett (1992) has estimated that between 100 and 200 hybrids are needed to map each chromosome, although James *et al.* (1994) used only 86, and a map of the whole genome would require over 4000 hybrids. This would be both expensive and laborious and would demand the availability of a complete set of human–rodent hybrids containing a single human chromosome. Walter *et al.* (1994) have found a solution to this problem by reverting to the original protocol of Goss and Harris (1975). Instead of using a human–rodent hybrid as a donor they used a diploid human fibroblast. Using 44 radiation hybrids they constructed a map of human chromosome 14 containing 400 ordered markers and concluded that a high resolution map of the whole human genome is feasible with only a single panel of 100–200 radiation hybrids.

Expressed sequence tags (ESTs)

In organisms with large amounts of repetitive DNA, the generation of an appropriate sequence and confirmation that it is an STS can be time consuming. Adams *et al.* (1991) have suggested an alternative approach. The principle of the method is based on the observation that spliced mRNA contains sequences that are largely free of repetitive DNA. Thus partial cDNA sequences, termed ESTs, can serve the same purpose as the random genomic STSs but have the added advantage of pointing directly to an expressed gene. In a test of this concept, partial DNA sequencing was conducted on 600 randomly selected human cDNA clones to generate ESTs. Of the sequences generated, 337 represented new genes, including 48 with similarity to genes from other organisms, and 36 matched previously sequenced human nuclear genes. Forty-six ESTs were mapped to chromosomes. In practice, ESTs need to be very short to ensure that the two ends of the sequence are contiguous in the genome, i.e. are not separated by an intron.

Polymorphic STSs

So far we have suggested that an STS yields the same product size from any DNA sample. However, STSs can also be developed for unique regions along the genome that vary in length from one individual to another. This variation in length most often occurs because of the presence of microsatellites. In man these usually take the form of CA (or GT) repeats with the dinucleotide being repeated 5–50 times. These sequences are very attractive because they are highly polymorphic, i.e. they will occur as $(CA)_{17}$ in one person, $(CA)_{15}$ in another, and so on. These repeat units are flanked by unique sequences (see Fig. 4.7) which can act as primers for the generation of an STS. By definition such STSs are polymorphic and can be traced through families along with other DNA markers.

Polymorphic STSs are particularly useful. First, because they occur on average every 10 kb they serve as very useful landmarks. Second, they act as landmarks on both the physical linkage map and the genetic linkage map for each chromosome and provide points of alignment between the different distance scales on these two types of maps. Weissenbach *et al.* (1992) used a total of 814 of such polymorphic STSs to produce a physical map of the human genome. By 1994, a total of 2066 polymorphic STSs had been mapped (Gyapay *et al.* 1994) with 56% of the genome at a distance of less than 1 cM (~1 Mb) from a marker. In a similar fashion, polymorphic STS maps have been constructed for the mouse (Dietrich *et al.* 1994), the cow (Barendse *et al.* 1994), the pig (Archibald 1994b) and the rat (Serikawa *et al.* 1992; Jacob *et al.* 1995). Polymorphic STSs also have been used extensively in plants (for review see Mazur & Tingey 1995) and may be combined with assays for the presence/absence of restriction enzyme sites to generate amplified fragment length polymorphisms (AFLPs).

Conventional autoradiographic procedures for analysing amplified microsatellites have limitations in large-scale projects requiring

CHAPTER 4
*Assembling a physical
map of the genome*

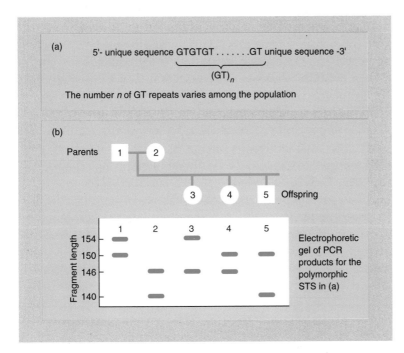

Figure 4.7 The use of a polymorphic STS in inheritance studies. (a) Structure of a polymorphic STS. (b) Schematic representation of a polymorphic STS in a five-member family. The two alleles carried by the father are different from those carried by the mother. The children inherit one allele of the STS from each parent. (Redrawn with permission from Dogget 1992 courtesy of University Science Books.)

hundreds of thousands of genotypes, particularly since the major source of error is in the reading of autoradiographs. There are, however, alternative methods. Diehl *et al.* (1990) used a laser-based automated DNA fragment analyser to size PCR products from dinucleotide repeats amplified using fluorescently labelled PCR primers. They were able to demonstrate the feasibility of this approach for large-scale human genetic mapping studies. Reed *et al.* (1994) now have produced a set of 254 markers covering all 22 human autosomes and the X chromosome. These markers proved to be highly efficient in the genotyping of microsatellites when incorporated in a standardized fluorescence-based protocol which used a software program to correlate the fluorescence intensity of labelled DNA fragments with marker loci information.

Disadvantages of STSs

Like restriction fragment mapping, STS mapping is still very labour intensive, even with extensive automation. In addition, it is not cheap: to synthesize two primers costs at least \$25–50. Thus the cost of generating 100–200 STSs is very significant. However, since gene synthesizers produce much more oligonucleotide than required, different laboratories are sharing their primers. Finally, an STS is defined by the PCR conditions that generate it. Unfortunately, many thermal cyclers used for PCR reactions are not accurately calibrated (Hoelzel 1990) making it difficult to reproduce exactly reaction conditions in different laboratories.

CHAPTER 4
*Assembling a physical
map of the genome*

RAPDs and CAPS

Polymorphic DNA can be detected by amplification in the absence of the target DNA sequence information used to generate STSs. Williams *et al.* (1990) have described a simple process, distinct from the PCR process, which is based on the amplification of genomic DNA with *single* primers of arbitrary nucleotide sequence. The nucleotide sequence of each primer was chosen within the constraints that the primer was nine or 10 nucleotides in length, between 50 and 80% G+C in composition and contained no palindromes. Not all the sequences amplified in this way are polymorphic but those that are (randomly amplified polymorphic DNA, RAPD) are easily identified. RAPDs are widely used by plant molecular biologists (Reiter *et al.* 1992; Tingey & Del Tufo 1993) to construct maps because they provide very large numbers of markers and are very easy to detect by agarose gel electrophoresis. However, they have two disadvantages. First, the amplification of a specific sequence is sensitive to PCR conditions, including template concentration, and hence it can be difficult to correlate results obtained by different research groups. For this reason, RAPDs may be converted to STSs (Kurata *et al.* 1994) after isolation. A second limitation of the RAPD method is that usually it cannot distinguish heterozygotes from one of the two homozygous genotypes.

A different method for detecting polymorphisms, which is not subject to the problems exhibited by RAPDs, has been described by Konieczny and Ausubel (1993). In this method, STSs are derived from genes which already have been mapped and sequenced. Where possible the primers used are chosen such that the PCR products include introns to maximize the possibility of finding polymorphisms. The primary PCR products are subjected to digestion with a panel of restriction endonucleases until a polymorphism is detected. Such markers are called CAPS (cleaved amplified polymorphic sequences). The way in which CAPS are detected is shown in Fig. 4.8. Note that whereas RFLPs are well suited to mapping newly cloned DNA sequences they are not convenient to use for mapping genes, such as plant genes, which are first identified by mutation. CAPS are much more useful in this respect.

Genome sequence sampling (GSS)

GSS is a recently described technique (Smith *et al.* 1994) which combines elements of restriction fragment mapping and STS mapping and generates maps with a resolution of 1–5 kb. As described above, STSs are used to prepare a physical map of YACs which ideally have been prepared from isolated chromosomes. To produce a high resolution map, a chromosome-specific cosmid library is prepared that represents the genome at 20–30-fold redundancy and contains

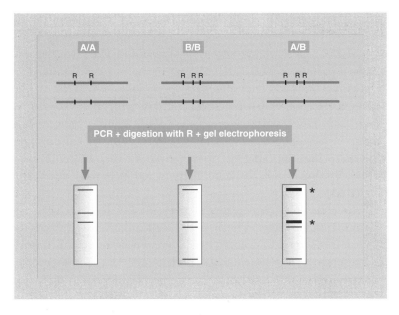

Figure 4.8 Generation and visualization of CAPS markers. Unique-sequence primers are used to amplify a mapped DNA sequence for two different homozygous strains (A/A and B/B) and from the heterozygote A/B. The amplified fragments from strains A/A and B/B contain two and three recognition sites, respectively, for endonuclease R. In the case of the heterozygote A/B, two different PCR products will be obtained, one of which is cleaved twice and the other three times. After fractionation by agarose gel electrophoresis the PCR products from the three strains give readily distinguishable patterns. The asterisks indicate bands that will appear as doublets. (Redrawn with permission from Konieczny & Ausubel 1993.)

a reasonably random distribution of clone ends. This is produced by cloning using a variety of restriction enzymes and cloning sites. Hybridization of YAC probes to the cosmid clones allows selection of large numbers of those included in the YAC clone. The cosmids are restriction mapped and arranged into contigs as described earlier. At the same time the restriction fragments covering the ends of each cloned piece of DNA are identified by hybridization with purified cosmid vector DNA. Next, automated DNA sequence analysis is carried out using cosmid DNA directly as template and primers recognizing each flanking region of the cosmid vector sequence contiguous to the insert. Thus the sequence of 300–500 bp of each end fragment can be determined with limited accuracy and aligned on the map. If a very high density cosmid map has been prepared then it enables a high density sequence map to be prepared. For example, Smith *et al.* (1994) studied the protozoan parasite *Giardia lamblia* which has a genome size of 10.5 Mb. A cosmid library of 20-fold redundancy would consist of 5000 unique cosmids and this would generate 10 000 end sequences. Assuming the restriction sites used for cloning were evenly spaced, the DNA sequences determined would be spaced every kilobase on average.

Hybridization mapping

This method starts with a genomic library as before. Five kinds of probes, representing known repetitive sequences (centromeric, telomeric, 17S and 5S ribosomal and the long terminal repeat (LTR) of retrotransposons), are hybridized to the library to identify those clones that contain only unique DNA. A number of clones carrying unique DNA are selected at random and used as hybridization probes to detect overlapping clones (see Fig. 4.9). From these clones which do not give a positive hybridization signal another set is selected at random for use as probes in the next round of experiments. This process is continued until all clones show positive hybridization at least once. In practice, some clones containing repetitive DNA have to be used to join contigs. Nevertheless, a key feature of this method is that clones are randomly chosen as probes based on only one criterion – that the clone has not yet given a positive hybridization signal. By this means, large numbers of redundant hybridizations are avoided.

Two refinements of the above process simplify the construction of contigs. First, probes can be prepared from either of the ends of the cloned DNA by using a vector with inward-facing T3 and T7 promoters located at the cloning site (Fig. 4.10). This simplifies contig generation compared with STSs because in the latter case it is hoped that the STS will lie in a region of overlap. However, if the ends of

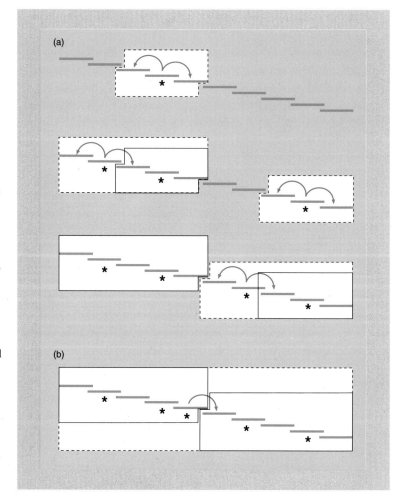

Figure 4.9 The principle of hybridization mapping. (a) Clones for use as probes are randomly picked (∗) from a given set of cosmids whose map order is not known. Hybridization identifies overlapping clones (arrows). From clones that do not give a positive signal in any earlier hybridization assay (unboxed areas), probes for the next round of experiments are chosen until all the clones show positive hybridization at least once. (b) Gaps in the map caused by the lack of probes for certain overlap regions are closed by using terminal contig clones. (Redrawn with permission from Hoheisel 1994.)

CHAPTER 4
*Assembling a physical
map of the genome*

the inserts were sequenced, rather than generating probes, then these sequences could be used as STSs. The use of insert ends as probes eliminates the problem of false positives (i.e. the presence of a hybridization signal between two clones that do not overlap) which can arise due to cross-hybridization between repetitive elements. This is done by demanding that all pairs of cosmid overlaps be reciprocal if both cosmids in the pair are used as probes (Zhang *et al.* 1994). That is, if cosmid x hybridizes to cosmid y then y must hybridize to x. If reciprocity is not achieved, that particular overlap is eliminated from the data set. Second, if a cosmid genomic library and a YAC genomic library are prepared from the same organism, the YAC library can be used to pre-sort the cosmid clones (Fig. 4.11). This mapping procedure can be refined still further, first by using as hybridization probes, short, random-sequence oligonucleotides (Fig. 4.11). Then, additional information can be obtained by using as probes sequences representing intron/exon boundaries, or zinc fingers or other structural motifs.

The utility of hybridization mapping has been shown by the construction of a map of the long arm of human chromosome 11 (Evans & Lewis 1989) and a complete map of the fission yeast (*Schizosaccharomyces pombe*) (Maier *et al.* 1992; Hoheisel *et al.* 1993; Mizukami *et al.* 1993).

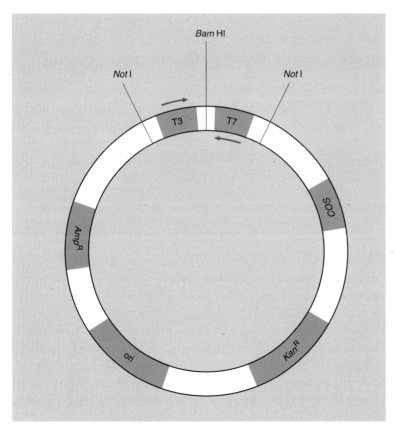

Figure 4.10 A cosmid vector used to generate probes specific for the ends of cloned inserts. The vector contains bacteriophage T3 and T7 promoters flanking a unique *Bam*HI cloning site. Note also the *Not*I sites to facilitate restriction mapping and excision of the insert DNA. ori, origin of replication; *Amp*^{R} and *Kan*^{R}, genes conferring resistance to ampicillin and kanamycin respectively; *COS*, cohesive sites essential for *in vitro* packaging in phage λ particles. Arrows show the direction of transcription from the T3 and T7 promoters.

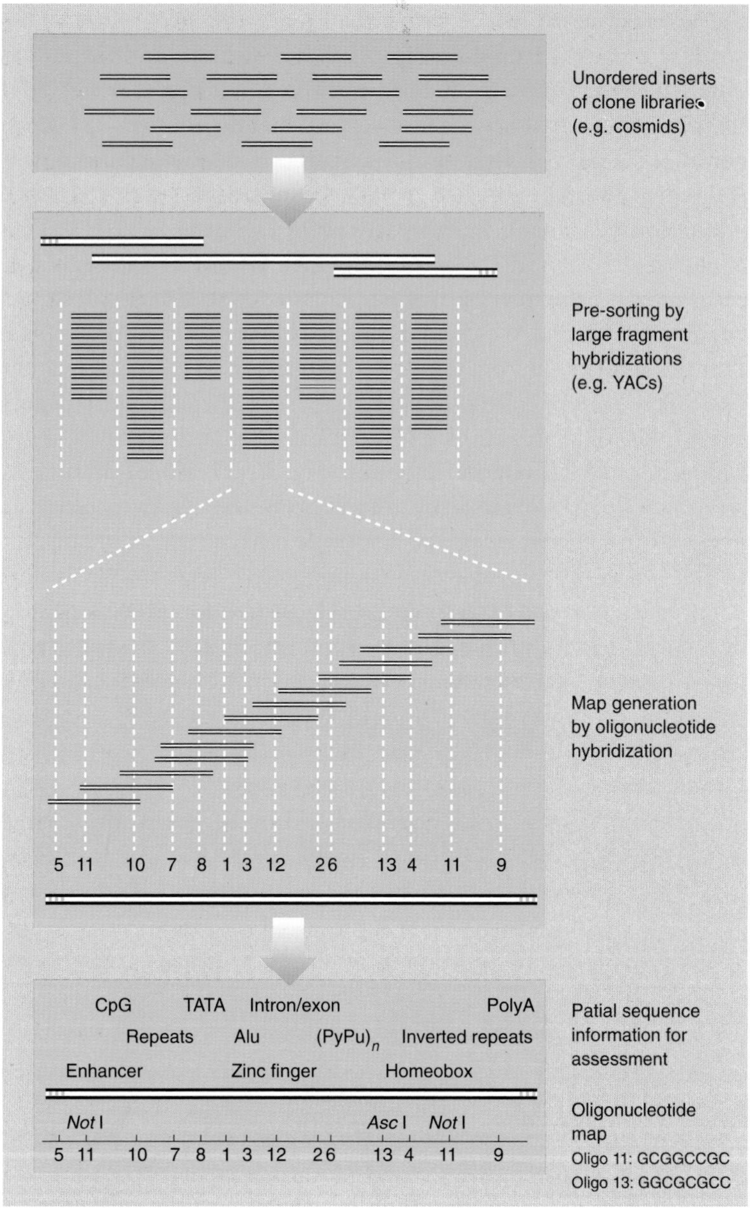

Figure 4.11 Principle of oligomer mapping. A high-resolution library (e.g. cosmids) is subdivided by hybridization of low-resolution DNA fragments (e.g. YAC clones). Fingerprinting data for establishing the order of clones are produced by hybridization with short oligonucleotides. Besides providing mapping data, this method yields an oligonucleotide map and partial sequence information concurrently. (Redrawn with permission from Hoheisel 1994.)

Labels in figure:

Unordered inserts of clone libraries (e.g. cosmids)

Pre-sorting by large fragment hybridizations (e.g. YACs)

Map generation by oligonucleotide hybridization

5 11 10 7 8 1 3 12 26 13 4 11 9

Patial sequence information for assessment

CpG TATA Intron/exon PolyA
Repeats Alu (PyPu)$_n$ Inverted repeats
Enhancer Zinc finger Homeobox

Oligonucleotide map

Not I *Asc* I *Not* I
5 11 10 7 8 1 3 12 26 13 4 11 9

Oligo 11: GCGGCCGC
Oligo 13: GGCGCGCC

Hybridization reference libraries

Once a genomic library has been prepared then, regardless of the vector used, it can be kept as a collection of individual clones in microtitre dishes. Each clone of the library then is spotted as a regular and reproducible array on a filter, ten or twenty thousand at a time. Batches of filters are distributed to other laboratories who screen them with probes generated by them from genes or regions of interest. Since a few standard-size filters containing an entire library can be probed multiple times, this is not a major undertaking. When positive hybridization signals are obtained, the location of the signal and the

CHAPTER 4
*Assembling a physical
map of the genome*

probe used are notified to the library holder who in turn provides a sample of the relevant clone(s) (Lehrach *et al.* 1990; Zehetner & Lehrach 1994). Thus one library can serve multiple users while at the same time enabling all data to be centralized to facilitate map generation (Fig. 4.12). Hybridization libraries are the antithesis of the STS approach: one uses clones as the reference objects, the other STSs. Similarly, one uses hybridization for screening, the other PCR. Whereas reproducible PCR conditions are essential for STS mapping, the library approach allows different hybridization conditions to be used by different groups.

The importance of *in situ* hybridization

There are a number of different kinds of genome maps, e.g. cytogenetic, linkage, physical, etc. The classic cytogenetic map gives visual reality to other maps and to the chromosome itself. Because it does not rely on the cloning of DNA fragments it avoids the pitfalls that this procedure can introduce, particularly with YACs (see p. 49). Genetic linkage mapping allows the localization of inherited markers relative to each other. As with cytogenetic maps, linkage maps examine chromosomes as they are in cells. Although the methodology used to construct cytogenetic and linkage maps can lead to errors, they nevertheless are used as gold standards against which the physical maps are judged. The importance of *in situ* hybridization is that it enables this comparison to be made. Providing hybridization of repeated sequences is suppressed and provided no DNA chimaeras are present, a cloned fragment or restriction fragment should anneal to a single location on the cytogenetic map. Furthermore, the physical map order should match that found by *in situ* hybridization. Where genetic markers have been located on the cytogenetic map by *in situ* hybridization they also can be positioned on the physical map.

Extensive cytogenetic maps exist for *Drosophila*, because of the presence of polytene chromosomes, and for man, because of the importance of clinical genetics. Thus it is not surprising that detailed *in situ* maps have been prepared for these two organisms (Lichter *et al.* 1990; Hartl *et al.* 1992). However, as well as being used to check physical maps, *in situ* hybridization can be used to generate physical maps. For example, in *Drosophila* YAC clones typically hybridize with enough polytene bands that overlapping clones (contigs) can be identified cytologically. In this context, cytological contigs are defined by the rule that two adjacent YACs are considered to overlap and if they have two or more polytene chromosome bands in common. The application of *in situ* hybridization to mapping in other organisms is much more difficult. Conventional fluorescent *in situ* hybridization (FISH) as applied to metaphase chromosomes has a resolving power of ~ 1 Mb. This is the upper size limit of YACs so will not allow fine mapping of cosmids. For higher resolution

CHAPTER 4
*Assembling a physical
map of the genome*

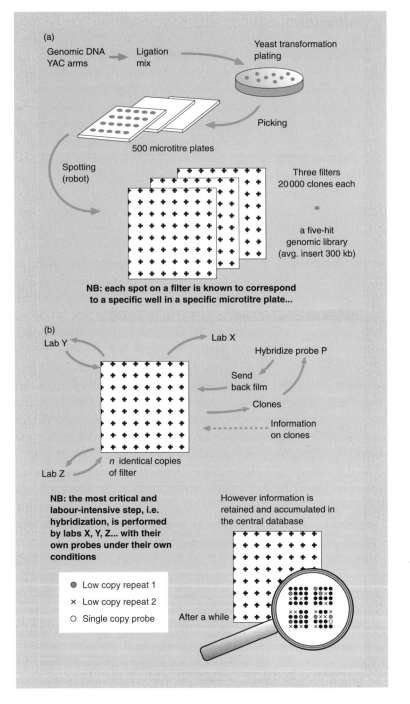

Figure 4.12 The concept of the reference library. (a) A YAC library containing about 50 000 clones stored in 500 microtitre plates. Each of these clones is individually and accurately positioned on the filters in a very fine grid (20 000 clones per filter) so that each corresponds to a defined well, in a particular microtitre plate. (b) The procedure for utilizing the reference library. Sets of identical filters are produced and sent to the laboratories wishing to isolate the YAC(s) corresponding to their probe(s). The laboratories then hybridize the filters and notify the central laboratory of the position of the positive clone(s) which can then be provided. The lower right half of (b) shows how information on each of the library's clones progressively and naturally builds up. (Redrawn with permission from Jordan 1993.)

mapping, FISH has been applied to interphase cell nuclei where the DNA is less condensed than in metaphase chromosomes. This permits resolution in the 50–100 kb range. Parra and Windle (1993) now have developed a technique for stretching DNA which allows even

74

greater resolution. By combining this procedure with differential fluorescent labelling of individual probes the arrangement of DNA probes along an extended strand of DNA can be observed. The images observed can be recorded through a fluorescence microscope and used to construct a direct visual hybridization (DIRVISH) DNA map.

As an example of DIRVISH mapping, consider two cosmid clones. Determining the relative orientation and distance between them by restriction mapping procedures would require the identification of a restriction fragment or fragments that connect the two sequences, or the isolation and mapping of overlapping clones. In such a test example Parra and Windle (1993) in two days were able to find the cytogenetic location of the two clones, show they were non-overlapping and that the gap between them was only 6 kb. DIRVISH also can be used to determine the relative orientation of two cosmids to facilitate chromosome walking. A DNA fragment from the end of one clone is hybridized in conjunction with the other clone. The distance between them indicates whether the fragment is from the proximal or distal end of the clone.

Computation and automation

Regardless of which method of physical mapping is used, the ordering of clones into contigs requires that an extensive amount of data has to be processed. Consider the experiments described on p. 57 for the mapping of the *E. coli* genome. A restriction enzyme fingerprint was prepared for 1025 clones and then all had to be compared pairwise to detect overlaps (see Fig. 4.2.). This is a mammoth task which is made even greater if the distance between restriction sites has to be included in the analysis as happened in the generation of the nematode and *Drosophila* maps. Thus it is not surprising that all the mapping methods make extensive use of computers and specially constructed algorithms. Over the years a number of different algorithms have been developed for contig generation using either restriction enzyme fingerprinting (Sulston *et al.* 1988; Branscomb *et al.* 1990) or hybridization methods (Mott *et al.* 1993; Wang *et al.* 1994; Zhang *et al.* 1994). The paper of Zhang *et al.* (1994) presents a particularly clear exposition of the principles of algorithm construction.

Much of the work involved in physical mapping is repetitive and monotonous. Surprisingly, there still is relatively little automation and the current status has been reviewed by Hodgson (1994). Whereas some steps are difficult to automate satisfactorily, e.g. isolation of ultrapure DNA, others such as Southern blotting are not. Similarly, some of the mapping techniques are compatible with automation, e.g. preparation of hybridization libraries, whereas others are not. Whenever possible, short cuts are used to reduce the number of repetitive steps. As just one example, consider hybridization mapping of cosmid clones described on p. 69. Starting from

CHAPTER 4
*Assembling a physical
map of the genome*

96 well archive plates, cosmid clones are inoculated on the surface of a filter. Each clone on the grid is assigned a unique identifying Y- and X-axis coordinate. To enable analysis of multiple clones simultaneously, cosmids are pooled according to the rows and columns of the matrix, DNA is prepared and a mixed RNA probe is synthesized. When hybridized to the matrix filter, the probe detects a pattern of spots consisting of all the template clones and the collection of clones overlapping with one end of each of the template clones. A similar procedure is carried out by using cosmids pooled according to columns of the matrix (Evans & Lewis 1989). When the two sets of data are compared, hybridizing clones identified by both of the mixed probes may be overlapping with the template clone common to both sets: the clone located at the intersection of the row and column (Fig. 4.13). It should be noted that with this method it is essential to eliminate hybridization due to repeat sequences. This is achieved by pre-hybridizing the cosmids with genomic DNA to low C_0t values (see p. 17).

A novel approach to generating maps: optical mapping

From the foregoing it should be clear that the construction of maps for eukaryotic chromosomes is laborious and difficult. Much of the problem derives from the fact that many of the procedures used for mapping and sequencing DNA were designed originally to analyse genes rather than genomes. The electrophoretic methods that are widely used in mapping offer the advantage of good size resolution, even for larger molecules, but require preparation of DNA in bulk amounts from sources such as genomic DNA or YACs. In contrast, single molecule techniques such as fluorescent *in situ* hybridization (FISH) make use of only a limited number of chromosomes but do not have good size resolution. Ideally, one would like to be able to combine the sizing power of electrophoresis with the intrinsic capability of FISH. Schwartz *et al.* (1993) have developed a technique, optical mapping, which approaches this ideal by imaging single DNA molecules during restriction enzyme digestion.

In practice, a fluid flow is used to stretch out fluorescently-stained DNA molecules dissolved in molten agarose and fix them in place during gelation. A restriction enzyme is added to the molten agarose–DNA mixture and cutting is triggered by the diffusion of magnesium ions into the gelled mixture which has been mounted on a microscope slide. Fluorescence microscopy is used to record at regular intervals cleavage sites which are visualized by the appearance of growing gaps in imaged molecules and bright condensed pools of DNA on the fragment end flanking the cut site. The size of the resulting individual restriction fragments is determined by relative fluorescence intensity and apparent molecular contour length. Wang *et al.* (1995b) have extended the technique to incorporate the use of RecA-assisted

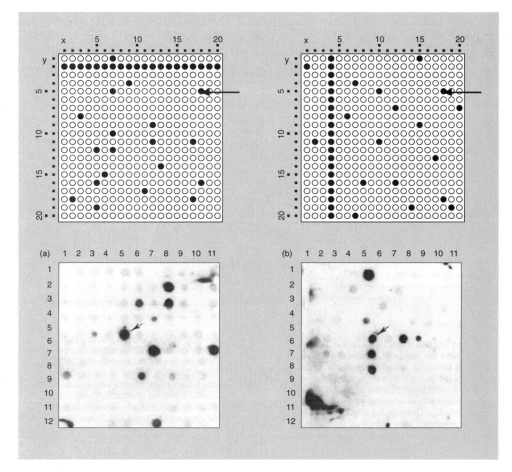

Figure 4.13 Strategy for analysis of physical linkage by using groups of cosmids (see text for details). (a) Schematic representation of the method. The arrows indicate the location of a clone that overlaps with the clone on coordinates y = 2, x = 4. (b) Actual results obtained using this procedure. (Reproduced with permission from Evans & Lewis 1989.)

restriction endonuclease (RARE) cleavage (see p. 40). Meng *et al.* (1995) have improved the size resolution of optical mapping and have accurately determined the mass of restriction fragments 800 bp long. To date, optical mapping has only been used in one laboratory and it remains to be seen if it will be more widely adopted. While it does not generate the same amount of mapping data as some of the other methods it does avoid the need for cloning. More importantly, it avoids the need for computation of the kind described in the previous section.

Table 4.1 Categorization of map objects (from Cox *et al*. 1994, with permission).

Mapping method	Experimental resource	Breakpoints	Markers
Meiotic	Pedigrees	Recombination sites	DNA polymorphisms
Radiation hybrid	Hybrid cell lines	Radiation-induced chromosome breaks	STSs
In situ hybridization	Chromosomes	Cytological landmarks	DNA probes
STS content	Library of clones	End points of clones	STSs
Clone-based finger-printing	Library of clones	End points of clones	Genomic restriction sites

Integration of different mapping methods and measuring progress

Each of the mapping methods described above has its advantages and disadvantages and so each method has its opponents and proponents. Thus it is common for different research groups to use different mapping methods even when working on the same genome. Ultimately the maps generated by the different methods need to be integrated. Also, it is essential to be able to monitor progress to distinguish those genomic regions requiring additional work and resources from those that essentially are complete. Cox *et al*. (1994) have provided a methodology for doing this.

Most genomic mapping projects involve ordering two classes of objects relative to one another. These are *breakpoints* and *markers* (Table 4.1). Breakpoints, so called because they represent subdivisions of the genome, are defined by a specific experimental resource. Markers consist of unique sites in the genome and should be independent of any particular experimental resource. Although both types of objects are essential for map construction, the map itself should be defined in terms of markers, especially those based on DNA sequence. One reason for this is that markers are permanent and easily shared. They can be readily stored as DNA sequence information and distributed in this fashion. By contrast, breakpoints are defined by experimental resources that tend to be transient and cumbersome to distribute. The most important reason to use sequence-based markers is that they can be easily screened against any DNA source and thus can be used to integrate maps constructed by diverse methods and investigators. Such integration is crucial to the assembly and assessment of maps and a good example is the comprehensive linkage map produced by three large, independent groups working on the human genome (Murray *et al*. 1994).

The breakpoints divide the genome into 'bins' corresponding to the regions between breakpoints. In assessing mapping progress a first step is to determine the number of 'bins' that are occupied with markers and the distribution of these markers within each bin. Although some investigators report only the total number of markers used to construct the map, it is the number of occupied bins that provides the measure of progress. The distribution of the number of markers per bin is important since the goal is to have markers evenly, or at least randomly, spread rather than clustered.

The second step in assessing progress is to identify those occupied bins that are ordered relative to one another and to estimate the confidence in the ordering. Note that assignment of markers to bins can proceed throughout a mapping project but the ordering of bins is only possible as a project matures. Thus ordering is a good indication of the degree of completion. Finally, the distance in kilobases between ordered markers in a map needs to be measured.

5 Sequencing methods and strategies

Basic DNA sequencing

The first significant DNA sequence to be obtained was that of the cohesive ends of phage λ DNA (Wu & Taylor 1971) which are only 12 bases long. The methodology used was derived from RNA sequencing and was not applicable to large-scale DNA sequencing. An improved method, plus and minus sequencing, was developed and used to sequence the 5386 bp phage ΦX 174 genome (Sanger *et al.* 1977a). This method was superseded in 1977 by two different methods, that of Maxam and Gilbert (1977) and the chain termination or dideoxy method (Sanger *et al.* 1977b). For a while the Maxam and Gilbert method, which makes use of chemical reagents to bring about base-specific cleavage of DNA, was the favoured procedure. However, refinements to the chain termination method meant that by the early 1980s it became the preferred procedure. To date, most large sequences have been determined using this technology with the notable exception of bacteriophage T7 (Dunn & Studier 1983). For this reason, only the chain termination method will be described here.

The chain terminator or dideoxy procedure for DNA sequencing capitalizes on two properties of DNA polymerases: (i) their ability to synthesize faithfully a complementary copy of a single-stranded DNA template, and (ii), their ability to use 2',3'-dideoxynucleotides as substrates (Fig. 5.1). Once the analogue is incorporated at the growing point of the DNA chain, the 3' end lacks a hydroxyl group and no longer is a substrate for chain elongation. Thus, the growing DNA chain is terminated, i.e. dideoxynucleotides act as chain terminators. In practice, the Klenow fragment of DNA polymerase is used because this lacks the 5'$\rightarrow$3' exonuclease activity associated with the intact enzyme. Initiation of DNA synthesis requires a primer and usually this is a chemically synthesized oligonucleotide which is annealed close to the sequence being analysed.

DNA synthesis is carried out in the presence of the four deoxynucleoside triphosphates, one or more of which is labelled with ^{32}P, and in four separate incubation mixes containing a low concentration of one each of the four dideoxynucleoside triphosphate analogues.

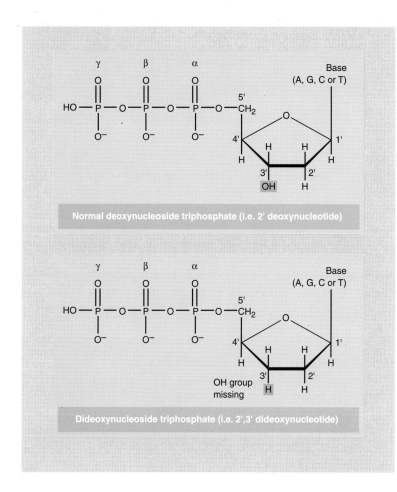

Figure 5.1 Dideoxy-
nucleoside triphosphates
act as chain terminators
because they lack a 3'-OH
group. Numbering of the
carbon atoms of the
pentose is shown (primes
distinguish these from
atoms in the bases). The
α, β and γ phosphorus
atoms are indicated.

Therefore, in each reaction there is a population of partially synthesized radioactive DNA molecules, each having a common 5'-end, but each varying in length to a base-specific 3'-end (Fig. 5.2). After a suitable incubation period, the DNA in each mixture is denatured and electrophoresed in a sequencing gel.

A sequencing gel is a high-resolution gel designed to fractionate single-stranded (denatured) DNA fragments on the basis of their size and which is capable of resolving fragments differing in length by a single base pair. They routinely contain 6–20% polyacrylamide and 7 M urea. The function of the urea is to minimize DNA secondary structure which affects electrophoretic mobility. The gel is run at sufficient power to heat up to about 70°C. This also minimizes DNA secondary structure. The labelled DNA bands obtained after such electrophoresis are revealed by autoradiography on large sheets of X-ray film and from these the sequence can be read (Fig. 5.3).

To facilitate the isolation of single strands the DNA to be sequenced may be cloned into one of the clustered cloning sites in the *lac* region of the M13 mp series of vectors (Fig. 5.4). A feature of these vectors is that cloning into the same region can be mediated by any one of a

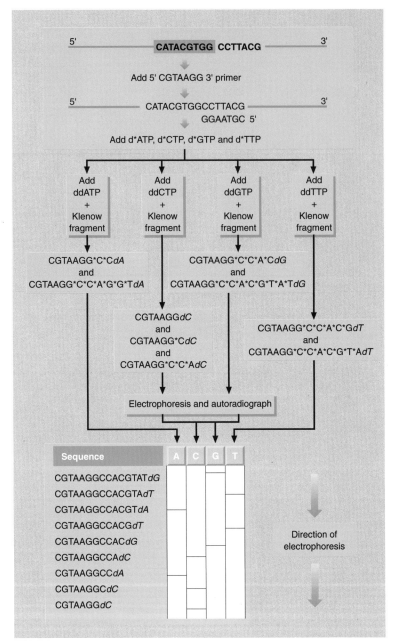

Figure 5.2 DNA sequencing with dideoxynucleoside triphosphates as chain terminators. In this figure asterisks indicate the presence of ^{32}P and the prefix 'd' indicates the presence of a dideoxy-nucleoside. At the top of the figure the DNA to be sequenced is enclosed within the box. Note that unless the primer is also labelled with a radioisotope the smallest band with the sequence CGTAAGG*dC* will not be detected by autoradiography as no labelled bases were incorporated.

large selection of restriction enzymes but still permits the use of a single sequencing primer.

Modifications of chain terminator sequencing

The sharpness of the autoradiographic images can be improved by replacing the ^{32}P-radiolabel with the much lower energy ^{33}P or ^{35}S. In the case of ^{35}S, this is achieved by including an α-^{35}S-deoxynucleoside triphosphate (Fig. 5.5) in the sequencing reaction.

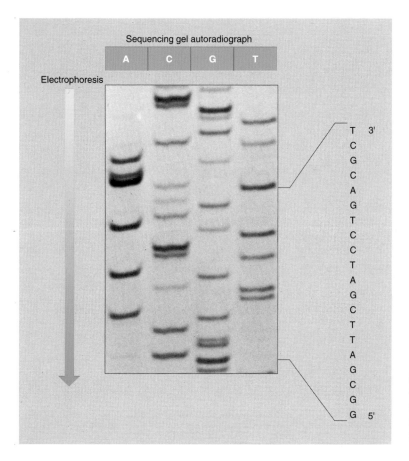

Sequencing gel autoradiograph

A C G T

Electrophoresis

T 3'
C
G
C
A
G
T
C
C
T
A
G
C
T
T
A
G
C
G
G 5'

Figure 5.3 Enlarged autoradiograph of a sequencing gel obtained with the chain terminator DNA sequencing method.

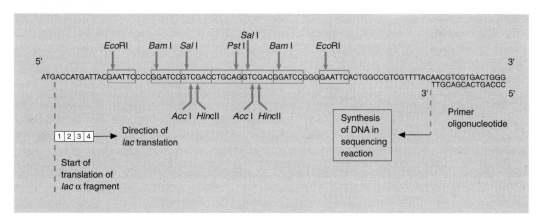

Figure 5.4 Sequence of M13 mp7 DNA in the vicinity of the multipurpose cloning region. The upper sequence is that of M13 mp7 from the ATG start codon of the β-galactosidase α-fragment, through the multipurpose cloning region, and back into the β-galactosidase gene. The short sequence at the right-hand side is that of the primer used to initiate DNA synthesis across the cloned insert. The numbered boxes correspond to the amino acids of the β-galactosidase fragment.

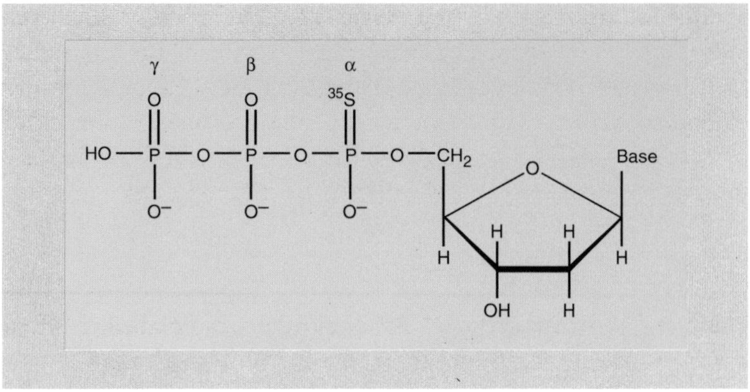

Figure 5.5 Structure of an
α-^{35}S-deoxynucleoside
triphosphate.

This modified nucleotide is accepted by DNA polymerase and incorporated into the growing DNA chain. Non-isotopic detection methods also have been developed with chemiluminescent, chromogenic or fluorogenic reporter systems. Although the sensitivity of these methods is not as great as with radiolabels, it is adequate for many purposes.

Other technical improvements to Sanger's original method have been made by replacing the Klenow fragment of *E. coli* DNA polymerase I. Natural or modified forms of the phage T7 DNA polymerase ('Sequenase') have found favour, as has the DNA polymerase of the thermophilic bacterium *Thermus aquaticus* (Taq DNA polymerase). The T7 DNA polymerase is more processive than Klenow polymerase, i.e. it is capable of polymerizing a longer run of nucleotides before releasing them from the template. Also, its incorporation of dideoxynucleotides is less affected by local nucleotide sequences and so the sequencing ladders comprise a series of bands with more even intensities. The Taq DNA polymerase can be used in a chain termination reaction carried out at high temperatures (65–70°C) and this minimizes chain termination artefacts caused by secondary structure in the DNA.

The combination of chain terminator sequencing and M13 vectors to produce single-stranded DNA is very powerful. Very good quality sequencing is obtainable with this technique, especially when the improvements given by ^{35}S-labelled precursors and T7 DNA polymerase are exploited. Further modifications allow sequencing of 'double-stranded' DNA, i.e. double-stranded input DNA is denatured by alkali, neutralized, and one strand then is annealed with a specific primer for the actual chain terminator sequencing reactions. This approach has gained in popularity as the convenience of having a universal primer has grown less important with the widespread availability of oligonucleotide synthesizers. With this development, Sanger sequencing was liberated from its attachment to the M13 cloning system. For example, PCR-amplified DNA segments can be sequenced directly. One variant of the double-

84

stranded approach, often employed in automated sequencing, is 'cycle sequencing'. This involves a *linear* amplification of the sequencing reaction using 25 cycles of denaturation, annealing of a specific primer to one strand only, and extension in the presence of Taq DNA polymerase plus labelled dideoxynucleotides. Alternatively, labelled primers can be used with unlabelled dideoxynucleotides.

Automated DNA sequencing

In manual sequencing, the DNA fragments are radiolabelled in four chain termination reactions, separated on the sequencing gel in four lanes, and detected by autoradiography. This approach is not well suited to automation. To automate the process it is desirable to acquire sequence data in real-time by detecting the DNA bands within the gel during the electrophoretic separation. However, this is not trivial as there are only about 10^{-15}–10^{-16} moles of DNA per band. The solution to the detection problem is to use fluorescence methods. In practice, the fluorescent tags are attached to the chain terminating nucleotides. Each of the four dideoxynucleotides carries a spectrally different fluorophore. The tag is incorporated into the DNA molecule by the DNA polymerase and accomplishes two operations in one step: it terminates synthesis and it attaches the fluorophore to the end of the molecule. Alternatively, fluorescent primers can be used with non-labelled dideoxynucleotides. By using four different fluorescent dyes it is possible to electrophorese all four chain terminating reactions together in one lane of a sequencing gel. The DNA bands are detected by their fluorescence as they electrophorese past a detector (Fig. 5.6). If the detector is made to scan horizontally across the base of a slab gel, many separate sequences can be scanned, one sequence per lane. Because the different fluorophores affect the mobility of fragments to different extents, sophisticated software is incorporated into the scanning step to ensure that bands are read in the correct order. A simpler method is to use only one fluorophore and to run the different chain terminating reactions in different lanes.

Automated DNA sequencers offer a number of advantages that are not particularly obvious. First, manual sequencing can generate excellent data but even in the best sequencing laboratories poor autoradiographs frequently are produced that make sequencing reading difficult or impossible. Usually the problem is related to the need to run different termination reactions in different tracks of the gel. Skilled DNA sequencers ignore bad sequencing tracks but many laboratories do not. This leads to poor quality sequence data. The use of a single-gel track for all four dideoxy reactions means that this problem is less acute in automated sequencing. Nevertheless, it is desirable to sequence a piece of DNA several times, and on both strands, to eliminate errors caused by technical problems. It should be noted that long runs of the same nucleotide or a high G+C content

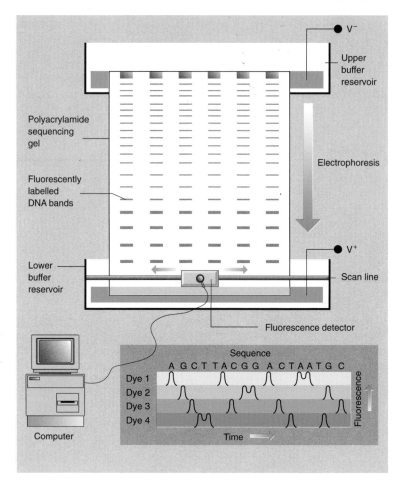

Figure 5.6 Block diagram of an automated DNA sequencer and idealized representation of the correspondence between fluorescence in a single electrophoresis lane and nucleotide sequence.

can cause compression of the bands on a gel necessitating manual reading of the data, even with an automated system. Note also that multiple, tandem short-repeats, which are common in the DNA of higher eukaryotes, can reduce the fidelity of DNA copying, particularly with Taq DNA polymerase. The second advantage of automated DNA sequencers is that the output from them is in machine-readable form. This eliminates the errors that arise when DNA sequences are read and transcribed manually.

The sequencing rate

The theoretical sequencing rate is easy to calculate. It is the number of sets of sequencing reactions that can be loaded on each gel, times the number of bases read from each sample, times the number of gels that can be run at once, times the number of days this can be done per year. For a 24-channel fluorescent sequencer this is about 2.7 million bases (Mb) per machine per year. Similar calculations can be made for manual sequencing. However, the best rate established during the sequencing of the 229 kb human cyto-megalovirus DNA was about 50 kb per person per year and the average was more like 25 kb (Barrell 1991)!

One way to increase the throughput of DNA sequencing is to use multiplexing (Church & Kieffer-Higgins 1988). This approach involves cloning the DNA to be sequenced into approximately 40 different vectors, each distinguished by a unique sequence that can be detected by probing a Southern blot of a sequencing gel. The different clones are mixed, subjected to the various sequencing reactions and the products are fractionated on a sequencing gel which is then blotted onto a membrane. The blot then is probed up to 40 times so that each time a different sequence will appear. Although the advantages of running fewer gels are offset by the need to do multiple probings and washings of each gel, the method does allow a real increase in speed.

Even with the development of multiplexing there are still three limitations of the basic Sanger method. The first of these is the time required to prepare the basic sequencing ladder and this probably cannot be improved without changing the basic sequencing chemistry. The second limitation is the length of time required to perform the electrophoretic separation. The simplest method for increasing separation speed is to increase the electrical potential but this significantly increases the amount of heat produced. Although sequencing gels are thin, they are sufficiently thick to develop a temperature gradient that leads to uneven migration rates. This causes the bands to become wider and ultimately reduces the resolution and hence the length of the sequence separated.

Two approaches have been developed to improve heat dissipation during electrophoresis. The first uses ultra-thin gels as the separation medium. By using thinner gels, heat is more rapidly diffused allowing the use of higher separation voltages (Kostichka *et al.* 1992). Because there are more uniform temperatures, the individual bands are more compact and this permits the use of less starting material. The second approach uses capillary tubes of gel instead of the usual slab gel. Using thin-walled capillaries provides for the rapid dissipation of the resistive heat generated at high voltages and permits separation time to be reduced by a factor of 25 (Luckey & Smith 1993). Arrays of capillaries could enable high throughput to be achieved especially when combined with improved detection methods, e.g. laser-excited confocal fluorescence scanning (Huang *et al.* 1992). A major technical problem with the use of capillaries is the difficulty in loading them.

An alternative approach to electrophoresis to separate the different sizes of sequencing fragments is matrix-assisted laser desorption ionization/time of flight/mass spectrometry (MALDI/TOF/MS). In this method, the DNA fragments are mixed with a carrier matrix and painted on the desorption target surface. A laser then is used to desorb and ionize the DNA fragments and acceleration in a mass spectrometer allows fragment length determination by time of flight detection. The biggest problem with this method is finding a suitable matrix from which to desorb the DNA (Lipshutz & Fodor 1994).

The third limitation of the dideoxy sequencing method is the gel resolving power for long DNA sequences. Gel separations of oligonucleotides 300–500 bp in length can be achieved routinely and under ideal conditions this can be extended to 1000 bp, i.e. 1 kilobase (kb). By contrast, the average bacterial genome is 5 Mb in length and the genome of higher eukaryotes can be more than 1000-times larger. The only way these large genomes can be sequenced is to sequence many short fragments and then join all these short sequences together. However, to assemble all these sequence fragments in order means that considerable sequence overlaps would be essential. Clearly, such a strategy is not feasible. How then can complete genome sequencing be undertaken?

Sequencing strategies

Two basic genome sequencing strategies have evolved. The first of these is complete genomic sequencing. Initially efforts were concentrated on viral or organelle genomes but more recently have extended to complete chromosomes (see below). This approach generates detailed and complete data but demands accuracy of sequence generation. It also is very labour intensive. As noted above, the maximum length of DNA that can routinely be sequenced in a single pass is 300–500 bp. Thus the sequencing of an entire chromosome means that many thousands of short overlapping sequences need to be determined. This, in fact, is the approach generally adopted but it demands the availability of a physical map so that the fragments sequenced can be re-assembled. It should be noted that with this approach sequencing is not the most difficult task. Rather, the construction of the physical maps is more demanding. So too is the handling of the vast amounts of data generated. As a rough guide, 5–10 kb of data are required to assemble 1 kb of confirmed sequence. The effort involved needs to be assessed in terms of the information derived from it. In prokaryotes and lower eukaryotes the gene density is very high (~ 900 genes per Mb) since introns are rare or absent and there is relatively little repetitive DNA. This is reduced to about 200 genes per Mb of DNA in the nematode (Wilson *et al.* 1994) and about 600 genes per Mb in yeast (Oliver *et al.* 1992). By contrast, a megabase of human DNA is expected to contain only 10 or 20 genes on average (Barrell 1991) although the gene density shows great regional variation.

An alternative sequencing strategy for higher eukaryotes is to analyse only cDNA. For example, in sequencing a highly spliced gene scattered in many exons over a megabase of human DNA, it would be much simpler to sequence the same coding region in 2 kb of cDNA rather than a megabase of genomic DNA. However, if the cDNA sequence is established first there is little incentive to sequence the genomic DNA at a later date! If one has only the genomic sequence then the cDNA sequence is needed to unambiguously identify the introns and exons. A variant of the cDNA sequencing approach has been adopted by a number of groups associated with the human genome project. In this, clones are selected at random from a cDNA library and a small amount of sequence determined for each, e.g. 200–300 nucleotides, a length that can be obtained in a single sequencing run. This information is incomplete since the full sequence of a cDNA usually comprises 1000–10 000 nucleotides. Furthermore, a low rate of errors in the sequence is not important for the sequence acts merely as a 'signature'. These signatures are used to scan databases to determine if they correspond to known genes. If so, no further work is necessary. If not, the entire cDNA sequence can be sequenced. Okubo *et al.* (1992) are using this approach with many different human cells and tissues in order to construct a 'body map of expressed human genes'.

While cDNA sequencing reduces the workload, e.g. expressed genes constitute only 5% of the human genome, it still is intellectually and technically demanding. The problems with this approach are threefold. First, control elements such as enhancers and promoters as well as splice sites would not be sequenced. Second, many genes are expressed at very low levels or for very short periods and so may not be represented in cDNA libraries. Finally, much of the DNA not sequenced, so-called 'junk' DNA, could have functions that are as yet unknown. Indeed, statistical analysis suggests that a language is contained within the 'junk' (Flam 1994). The chromosome is an organelle that self-replicates and transcribes information. It is physically manoeuvred during mitosis and meiosis and it participates in recombination. The structure of the chromosome also appears to play a role in regional mutation rates and therefore affects the evolutionary process. It is likely that as yet uncharacterized DNA sequences are involved in these chromosomal functions and hence the whole genome ultimately needs to be sequenced. There also is some debate as to what actually constitutes a gene and therefore there is no agreement as to how many genes there are in each genome (for discussion see Fields *et al.* 1994; Bird 1995).

Krishnan *et al.* (1995) recently have suggested a third sequencing strategy called *feature mapping*. This is used to help guide the choice of regions to be subjected to detailed analyses.

Genomic sequencing

The phage, cosmid or YAC clones used in the preparation of physical maps are too large to be sequenced directly. Consequently, clones are subjected to random subcloning prior to sequencing. Early in the project this results in rapid accumulation of sequence data but later is subjected to the law of diminishing returns. At a point where approximately 95% of a sequence has been determined a switch is made to a directed approach to obtain the remainder of the sequence. Note that for this strategy cosmids are the vector of choice because a large amount of DNA can be purified (cf. YACs) and the ratio of insert:vector sequence is high (9:1), cf. phage vectors. The detailed procedure described here is that of Wilson *et al.* (1992). All genomic sequencing groups follow a similar strategy although the detailed tactics may vary. In particular, many groups are moving to the use of cycle sequencing (p. 84).

The first step is to fragment the cloned DNA in a random manner and sonication has proved to be the most effective method. After sonication, fragments in the size range 0.8–1.6 kb are selected by preparative agarose gel electrophoresis and subcloned in a suitable M13-based sequencing vector. Smaller fragments are less useful for sequencing whereas larger fragments can undergo deletion during propagation in M13. The next step is to sequence random subclones and this is done using the M13 universal primer. For accuracy the same sequence needs to be obtained from 5–6 different clones and should be obtained for both strands. The number of different, randomly-selected subclones that need to be sequenced to cover 95% of the original cosmid clone depends on the size of that primary clone and can be determined using the statistical profiles of Bankier and Barrell (1983). After DNA sequence analysis of the appropriate number of random subclones, sequence contigs are assembled using appropriate computer programs. Gaps in the contigs have to be closed and this is done by synthesizing PCR primers homologous to the

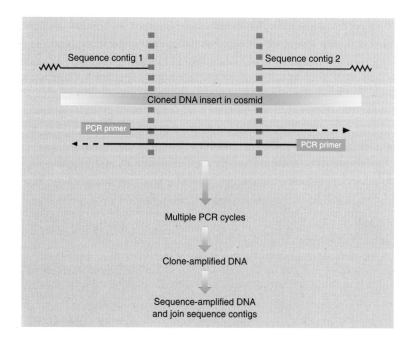

Figure 5.7 Linking DNA sequence contigs by walking.

ends of the contigs and using them to isolate the missing sequence (Fig. 5.7). This primer directed 'walking strategy' is continued until all the gaps have been closed.

The report of Wilson *et al.* (1992) provides detailed information on the logistics of sequencing a large piece of DNA, in this case, 96 kb of mouse DNA. Initially they used radioisotopic methods and a relatively high throughput was maintained using robotics to perform DNA sequencing reactions and to load gels. Up to 24 subclones could be processed per person per day and another full day was required to process the data generated. The maximum throughput obtained was 96 subclones per person per week. To reduce the burden of manual data entry from autoradiographs they switched to automated sequencing instruments based on fluorescent dye chemistry. Under optimal conditions it was possible to process up to 120 subclones per instrument per week. The magnitude of the task can be appreciated when it is realized that Wilson *et al.* (1992) sequenced 2399 subclones in order to generate their 96 kb of mouse DNA sequence! A significant fraction (27.5%) of these subclones did not generate useful sequence because they either lacked an insert or a primer site or could not be produced in sufficient amounts for sequence analysis. In addition, using the automated analyser, instrument failures and downtime are significant (15–20%). It remains to be seen if some of the more recent improvements (p. 87) to the basic Sanger sequencing method can significantly improve the throughput in large-scale sequencing procedures of this kind.

A number of the subclones generated in the above sequencing procedure will contain fragments of the cosmid vector. Some groups, e.g. Edwards *et al.* (1990), have described the identification and exclusion of such subclones using plaque hybridization. Wilson *et al.* (1992) chose not to direct any efforts into pre-screening but to process the relatively small number (12%) of subclones containing regions of the cosmid vector. Not only did it eliminate tedious plaque hybridization but it allowed error analysis by comparison of raw sequence data with known vector sequences. Sequences that were read beyond 400 bp contained an average of 3.2% error while those less than 400 bp had 2.8% error. At least one-third of the errors were due to ambiguities in sequence reading. In those sequences longer than 400 bp that were read, most errors occurred late in the sequence and often were present as extra bases in a run of two or more of the same nucleotide. The remainder of the errors were due to secondary structure in the template DNA. However, because the complete sequence was analysed with an average 5.9-fold redundancy and most of it on both strands, the final error frequency is estimated to be less than 0.1%. By comparison, 35 different European laboratories are engaged in sequencing the *S. cerevisiae* genome with the attendant possibility of a very high error frequency. However, by using a DNA coordinator who implements quality control procedures

Method of verification	Total number of fragments	Total bp verified	Error % detected
Original overlap between cosmids	28	63 424	0.02
Resequencing of selected segments (3–5 kb long)	21	72 270	0.03
Resequencing of random segments (~ 300 bp long)	71	18 778	0.05
Resequencing of suspected segments (~ 300 bp long) from designed oligonucleotide pairs	60	17 035	0.03
Total	180	171 507	
Average error rate			0.03

(Table 5.1), the overall sequence accuracy for yeast chromosome XI (666 448 bp) was estimated to be 99.97% (Dujon *et al.* 1994), i.e. similar to that noted above. Lipshutz *et al.* (1994) have described a software program for estimating DNA sequence confidence. Fabret *et al.* (1995) have analysed the errors in finished sequences. They took advantage of the fact that the surfactin operon of *Bacillus subtilis* had been sequenced by three independent groups. This enabled the *actual* error rate to be calculated. It was found to range from 0.02 to 0.27%, the different error rates being ascribed to the detailed sequencing tactics used.

DNA sequence databases

Since the current DNA sequencing technology was developed a large amount of DNA sequence data have accumulated. These data are maintained in three databases: GenBank in the USA (Benson *et al.* 1994), the DNA Database of Japan, and the European Bioinformatics Institute (EBI) database in Europe (Emmert *et al.* 1994) which has replaced the original EMBL database (Rice *et al.* 1993). Each of these three groups collects a portion of the total sequence data reported world wide. All new and updated database entries are exchanged between the groups on a daily basis. Users world wide can access the databases via electronic mail.

The rate of growth of the databases is accelerating. The June 1994 version of the EBI database contained 200 million bases from 183 thousand entries and was 70% larger than one year previously. The amount of data accumulated for the major genomes under study is shown in Table 5.2. Note that not all the sequence information deposited is unique. For example, by the end of June 1994, 4.6×10^6 bp of *E. coli* sequence had been deposited but only 2.9×10^6 bp of unique sequence, i.e. 61% of the entire genome (Wahl *et al.* 1994).

Organism	Genome size (Mb)	Database size (Mb)	Unique fraction (Mb)	% of genome sequenced
E. coli	4.7	5.93	3.69	78.5
Bacillus subtilis	4.2	1.56	1.14	27.1
S. cerevisiae	15	8.12	5.29	35.3
S. pombe	14	0.85	0.72	5.1
C. elegans	90	4.7	4.43	4.9
A. thaliana	100	1.38	1.17	1.2
D. melanogaster	170	5.62	4.02	2.4

Table 5.2 Progress towards a complete sequence of some genomes. Data on deposit at the European Bioinformatics Institute as of November 1994.

Analysing sequence data

Contemporaneously with the increase in sequence data have been rapid increases in the power of computers and improvements in software. Thus, data from genome sequencing can be scanned quickly to identify start and stop signals for transcription, sequence repeats, Z-DNA and restriction enzyme recognition sites (Rice *et al.* 1991). The latter are particularly useful for they enable a cross-check to be made with any physical map generated using restriction endonucleases (see pp. 57 *et seq.*). More important, the data generated in a genome sequencing project can be compared with the sequences generated from individual gene cloning projects. For example, 725 kb of contiguous sequence have been generated from *E. coli* (Sofia *et al.* 1994) and this represents 15.5% of the genome yet over 78% of the genome is represented in the central databases (Table 5.2). The kind of information that can be generated from genome sequencing projects is shown in Table 5.3. It is worth noting the different gene densities in the different genomes (Table 5.4).

Uberbacher and Mural (1991) have developed a reliable computational approach for locating protein-coding portions of genes in anonymous DNA sequences. Their method combines a set of sensor algorithms and a neural network to localize the coding regions. From the nucleotide sequence of a gene it is easy to deduce the protein sequence that it encodes. Unfortunately, nobody yet has formulated a set of general rules that allows us to predict a protein's three-dimensional structure from the amino acid sequence of its polypeptide chain. However, based on crystallographic data from over 300 proteins, certain structural motifs can be predicted. Nor does an amino acid sequence on its own give any clue as to function. The solution is to compare the amino acid sequence with that of better-characterized proteins: a high degree of homology suggests similarity in function. In this way Koonin *et al.* (1994) analysed the information contained in the complete sequence of yeast chromosome III and found that 61% of the probable gene products had significant similarities to sequences in the current databases. As many as 54% of them had known functions or were

Table 5.3 Selected data abstracted from major genome sequencing projects.

Reference	Organism and length of DNA sequenced	Key features of genome discovered from sequence data
Sofia *et al.* (1994)	*Escherichia coli* (225.4 kb)	191 putative coding genes (ORFs) of which 72 previously known, 9 identified and 110 unidentified despite similarity searches. Arrangement of possible promoters and terminators suggests 90 transcription units. 31 putative signal peptides found including 13 known membrane or periplasmic proteins. Also found were 1 tRNA gene, 2 insertion sequences, 19 REP elements, 50 Chi sites and 95 computer-predicted DNA bends
Oliver *et al.* (1992)	*Saccharomyces cerevisiae* chromosome III (315 kb)	182 ORFs for proteins longer than 100 amino acids, of which 37 previously known and 29 show some similarity to sequences in databases. Three genes were found to contain introns. 10 tRNA genes were identified and all were within 500 bp of a Ty or delta element. The sequence data also provide evidence of post-recombination events
Dujon *et al.* (1994)	*Saccharomyces cerevisiae* chromosome XI (666.4 kb)	331 ORFs for proteins longer than 100 amino acids, of which 93 correspond to identified genes. Of the remainder, 93 have homologues among gene products of yeast or other organisms and 37 have homologues of unknown function. 16 tRNA genes were found, 3 with introns, and 1 sigma and 11 delta sequences. The % G + C content varies along the genome and correlates with gene density
Johnston *et al.* (1994)	*Saccharomyces cerevisiae* chromosome VIII (562.6 kb)	269 ORFs for proteins longer than 100 amino acids of which 59 were previously identified. Of the 210 novel genes, 65 are predicted to encode proteins with similarity to other known proteins. 11 tRNA genes were identified, 3 with introns. Coding density and base composition across the chromosome are not uniform but no regular pattern of variation apparent.
Feldmann *et al.* (1994)	*Saccharomyces cerevisiae* chromosome II (807.2 kb)	General features as for other yeast chromosomes. Functional ARS elements, tRNA genes, and Ty elements preferably located in AT-rich regions. High degree of internal genetic redundancy
Wilson *et al.* (1994)	*Caenorhabditis elegans* (2.2 Mb)	483 putative genes identified. 29% of sequenced region is coding sequence with 48% representing putative introns and exons. Gene density is one gene per 4.5 kb. No overlapping genes. 14 tRNA genes identified, 3 within introns of likely genes. Inverted repeats found on average every 5.5 kb, tandem repeats on average every 10 kb. Interspersed repeats found with an average of 14 copies per 100 kb
Wilson *et al.* (1992)	Mouse immunoglobulin locus (94.6 kb)	15 novel TcrαJ gene segments discovered, all randomly dispersed and oriented in same transcriptional direction. Pseudogenes also detected. Information provided concerning the organization and structure of T-cell receptor genes and how they rearrange in lymphoid cells
Martin-Gallardo *et al.* (1992)	Human chromosome 19q13.3 (106 kb)	5 genes found, of which 2 are of unknown function. Has average density of 1.4 *Alu* repeats per kilobase
McCombie *et al.* (1992)	Human chromosome 4p16.3 (58 kb)	Sequenced region containing 62 *Alu* repeats, 11 exons and 3 expressed genes for unidentified proteins. Each gene associated with CpG island. *Alu* repeats have average length of 300 bases. Sequencing of 31 kb DNA from 2 individuals revealed 72 DNA sequence polymorphisms

continued

Table 5.3 (*Continued*)

Reference	Organism and length of DNA sequenced	Key features of genome discovered from sequence data
Iris *et al.* (1993)	Short arm human chromosome 6 (90 kb)	Sequenced region contains 4 known genes and a novel telomeric potential coding region. The genes are bracketed by long, dense clusters of *Alu* repeats belonging to all the major families. At least 6 new families of medium reiteracy frequency repeats (MERS) and one pseudogene are intercalated within and between the *Alu* clusters
Koop & Hood (1994)	Human and mouse T-cell receptor DNA	Comparative DNA sequence analysis showed a very high level (71%) of organizational and coding and non-coding sequence similarity

Organism	Number of protein-coding genes per 100 kb sequence
E. coli	87
S. cerevisiae	52
C. elegans	22
Human	5

Table 5.4 Gene densities in different organisms determined from large-scale sequencing projects (Data abstracted from Table 5.3).

related to functionally characterized proteins and 19% were similar to proteins of known three-dimensional structure. Other examples are given in Table 5.3.

Because the DNA and protein sequence databases are updated daily it could be that homologues of previously unidentified proteins have been found. Thus it could pay to re-search the databases at regular intervals as Robinson *et al.* (1994) found recently. Starting with more than 18 million bp of prokaryotic DNA sequence they executed a systematic search for previously undetected protein-coding genes. They removed all DNA regions known to encode proteins or structural RNAs and used an algorithm to translate the remaining DNA in all six reading frames. A search of the resultant translations against the protein sequence databases uncovered more than 450 genes which previously had escaped detection. Seven of these genes belonged to gene families not previously identified in prokaryotes. Others belonged to gene families with critical roles in metabolism. Clearly, periodic exhaustive reappraisal of databases can be very productive!

If homology with known proteins is to be used to analyse sequence data then sequence accuracy is paramount. With the gene density and ORF size distribution of yeast (Dujon *et al.* 1994) and the nematode (Wilson *et al.* 1994), even relatively rare sequencing errors, many of which are missing nucleotides, will result in a large fraction of the protein coding genes being affected by frameshifts. At 99%

accuracy virtually all genes will contain errors and at 99.9% accuracy two-thirds of the genes will still contain an error, most likely a frameshift. With the 99.97% accuracy obtained in the yeast sequencing about one-third of predicted genes will still contain errors.

Because sequence similarity search programs are so vital to the analysis of DNA sequence data, much attention has been paid to the precise algorithms used and their precise speed. However, the effectiveness of these searches is dependent on a number of factors such as the choice of scoring system, the statistical significance of alignments and the nature and extent of sequence redundancy in the databases (Altschul *et al.* 1994; Coulson 1994). Most database search algorithms rank alignments by a score whose calculation is dependent upon a particular scoring system. Optimal strategies for detecting similarities between DNA protein-coding regions differ from those for non-coding regions. Special scoring systems have been developed for detecting frameshift errors in the databases, a problem highlighted above. Thus a database search program should make use of a variety of scoring systems. Furthermore, given a query sequence, most database search programs will produce an ordered list of imperfectly matching database similarities, but none of them need have any biological significance. An important question is how strong a similarity is necessary to be considered significant and a number of procedures now are available for assessing database search results.

Surprisingly strong biases exist in protein and nucleic acid sequences and sequence databases. Many of these reflect fundamental mosaic properties that are of interest in their own right. Databases also contain some very large families of related domains, motifs or repeated sequences, in some cases with hundreds of members. In other cases there has been a historical bias in the molecules that have been chosen for sequencing. Unless special measures are taken, these biases very commonly confound database search methods and interfere with the discovery of interesting new sequence similarities.

The completeness of the database examined is as important as the speed and sensitivity of the search. The main sequence databases are published on a quarterly basis although all can provide release of new entries on a daily or weekly basis. However, the efforts needed to keep local copies of databases up to date can be prohibitive and hence it may be necessary for the sequence generated to be remotely searched at a specialized centre. Remote services are commonly provided by electronic mail. The user sends the sequence and parameters of the search by electronic mail to the specialized centre who analyse the data and return the output, again by electronic mail. However, the possibilities for interacting with the search program are limited. For most users, the primary consideration in carrying out a database search is the timeliness and completeness of the database searched. There is much greater chance of missing an important feature by using a poorer program on a better database than the reverse!

cDNA sequencing

As noted earlier (p. 88), cDNA sequencing represents a complementary approach to complete genome sequencing and is particularly appropriate for use with higher eukaryotes. Whereas complete genome sequencing yields information on the physical structure of the genome, cDNA sequencing can give information on gene expression, e.g. what genes are expressed, and to what extent, in any given cell or tissue at a particular time.

There are several different strategies for constructing a cDNA library that can greatly influence the available set of information. First, a cDNA library may be prepared from a selected cell or organ so the library constituents reflect the physiology of that cell or organ (Weinstock *et al.* 1994). Alternatively, combining several cDNA sub-libraries from various tissues maximizes the number of gene transcripts. Second, the library may be designed so as to represent faithfully the abundance of gene transcripts in the original mRNA population or, alternatively, to represent a collection of non-overlapping transcripts. The latter type, a *single book* library, is used for surveying as many gene transcripts as possible. The generation of a single book library requires the use of a subtractive cDNA cloning procedure, preferably one that needs small amounts of starting mRNA (Travis & Sutcliffe 1988) or a cloning procedure which specifically generates a nearly equal representation of cDNAs (Ko 1990). Without the use of such procedures, low abundance mRNAs are unlikely to be cloned as cDNAs. Finally, the library may be designed to carry full-length inserts or, alternatively, it may consist of partial cDNAs covering a directed portion of each mRNA, such as the 3' or 5' ends. Composition of the latter libraries can be proportional to the composition of the mRNA population, in constrast to the full-size library. Sequencing from the 3' end of cDNA clones generates mainly 3' untranslated sequence which is helpful in discriminating between homologous gene family members but gives little information about gene function. It also results in repeated sequencing of uninformative poly(dA) stretches despite the use of oligo(dT) primers. Although 5' sequencing can overcome these limitations, varying clone lengths mean that sequence information may not be generated from the same position in homologous clones. Thus true identity may be missed.

Using a random selection approach, Adams *et al.* (1991) found an unacceptably large number of highly represented clones in their cDNA libraries. Over 30% of clones from human hippocampal DNA were uninformative. Nevertheless, from 600 clones sequenced, 337 represented new genes. In an extension of this work, Adams *et al.* (1992) sequenced an additional 2672 independent clones and identified 2375 new sequences. To eliminate highly represented cDNAs, Khan *et al.* (1992) pre-screened their library with labelled

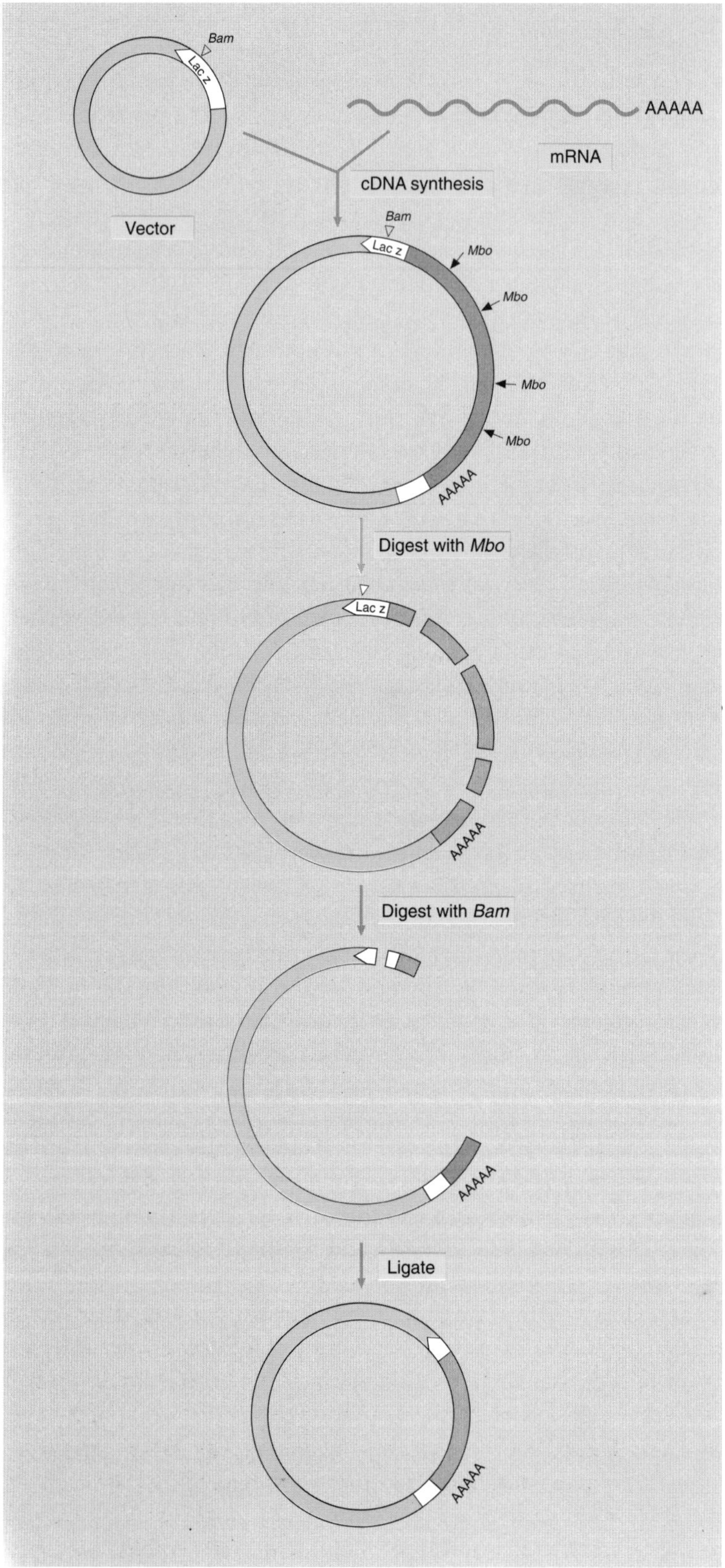

Figure 5.8 The method used to clone the 3' end of a cDNA. The sequencing vector DNA is methylated prior to cloning the cDNA to make it resistant to endonuclease *Mbo*I. This can be achieved by *in vitro* methylation or by preparing the vector from a *dam*, host. The mRNA is converted to cDNA using the Okayama and Berg procedure (see Old & Primrose 1994). The clones carrying the cDNA are digested with *Mbo*I to leave only the 5' and 3' ends of the cDNA. The 5' end then is removed by digestion with *Bam*HI. Note that the overhang produced by endonucleases *Mbo*I and *Bam*HI are identical.

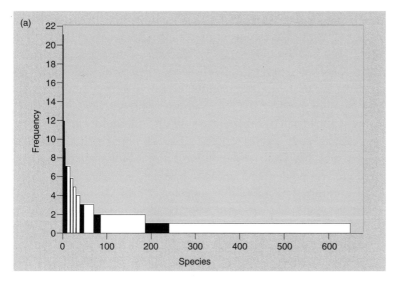

(a)

(b)

Frequency (in 982)	Representative clone	Insert size (nt)	Gene
22	a29	523	Elongation factor in 1α
21	c12a04	132	Serum albumin
12	a26	188	Translationally controlled tumour protein
9	c12a01	261	Unknown
8	a31	369	α-1-antitrypsin
7	a23	183	Rat ribosomal protein L21
7	c12b03	25	Ribosomal protein L31
7	c12b06	88	Rat ribosomal protein S19
7	c13c04	174	Mouse ribosomal protein L3
7	d2g03	101	Unknown
6	a-30	332	Ferritin light chain
6	a-31	156	Unknown
6	a20	383	Unknown
6	d0g12	107	Unknown
6	hm05a02	373	Unknown
6	s304	120	Unknown
5	a29	182	Ribosomal protein S16
5	a32	190	Apolipoprotein A11
5	hm02g10	163	Unknown
5	s406	364	Unknown

Figure 5.9 Expression profiles of genes in a human liver cell line. The frequencies of the appearance of each cDNA species in 982 randomly sequenced clones are shown in (a). Species are arranged in the order of their frequencies of appearance. Solid and open bars represent the novel and identified sequences respectively. (b) This table shows some properties of the most abundant 20 species. (Redrawn with permission from Okubo *et al.* 1992 courtesy of *Nature Genetics.*)

total brain cDNA and an excess of cold genomic DNA. Plaques that were not labelled were selected and gave a sevenfold increase in the number of new sequences identified. Of 1024 human brain cDNAs, over 900 represented new human genes. Okubo *et al.* (1992) used as the basis of their sequencing a 3'-directed cDNA library (Fig. 5.8) from a liver cell line which is a non-biased representation of the mRNA population. In total, 982 random cDNA clones were sequenced and, because of the random selection of clones, several identical sequences emerged. The frequency distribution of these is shown in Fig. 5.9. Altogether, among the 982 clones that produced informative sequence data, 215 clones represented 91 known genes and 767 clones represented 550 novel genes. Adams *et al.* (1993a,b) have reported the sequencing of yet another 5000 cDNAs and once again many interesting genes have been identified. Whereas Adams *et al.* (1993a,b) have concentrated on brain cDNAs, Okubo *et al.* (1992) studied liver cDNAs. Comparison of the sequence profiles from these two tissues indicates that structural and regulatory proteins are more abundant in the brain but secreted proteins and polypeptides involved in protein synthesis predominate in the liver.

In Okubo *et al.* (1992), their work with 982 random cDNA clones generated over 270 kb of sequence data. As noted earlier (p. 91) the accuracy of this sequence data is important. If the sequence data are to be used for predicting translation products, sequencing errors need to be minimized by multiple sequencing efforts. On the other hand, if the sequence data are used only for identifying mRNA species, handling a large number of samples is more important than the accuracy of the sequence data. In all the studies to date, the number of mRNA species identified has been the key criterion and single pass sequencing has been used.

A logical extension of single pass sequencing of cDNAs is determination of the precise chromosomal location of each of the corresponding genes. While cDNAs can be used directly for FISH mapping to metaphase chromosomes, this can be technically difficult. Also, while regional mapping information in generated, no physical genomic clone corresponding to the cDNA is produced. However, mapping cDNAs to genomic cosmids or YACs and then using the latter as FISH probes provides both regional mapping data and a cloned piece of genomic DNA from which the cDNA is derived. Polymeropoulos *et al.* (1993) have mapped more than 300 human brain ESTs. Surprisingly, their distribution did not correlate with the cytological length of the chromosomes. For example, chromosomes 2 and 19 contain similar numbers of genes despite their large difference in size whereas chromosomes 13 and 18 were relatively under-represented. The latter observation may explain why the only viable human trisomies are of chromosomes 13, 18 and 21.

In the vast majority of cases, the demonstration of sequence identity between a newly obtained EST and a previously described gene will be correctly interpreted as evidence for the expression of that specific gene in the tissue from which the EST was derived. However, Tsai *et al.* (1994) have shown that this is not always the case. They showed that a gene called *TCTE1* was not expressed in tissues other than the male reproductive track whereas Adams *et al.* (1992, 1993a,b) reported putative *TCTE1* ESTs in cDNA clones from human brain libraries. However, detailed analysis showed that EST-defined clones with homology to *TCTE1* sequences are derived from a transcription unit present on the opposite strand of *TCTE1*.

The use of cDNA sequencing is not restricted to human DNA: sequencing projects are underway in *Caenorhabditis*, *Arabidopsis*, *Oryza sativa* (rice), *Plasmodium falciparum* (malaria), *Drosophila* and the mouse (Sikela & Auffray 1993; Boguski *et al.* 1993). Data

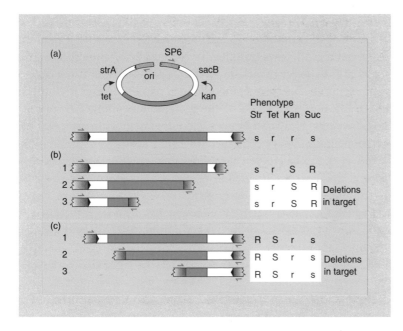

Figure 5.10 Schematic representation of hypothetical nested deletions obtained using a transposon-based cosmid vector. (a) Depiction of a hypothetical DNA clone. The green portion represents cloned DNA, the filled arrowheads the transposon sequences and the green half-arrows represent locations of primer binding sites. (b) Deletions extending into various sites within the cloned fragment resulting from selection for resistance to sucrose, caused by loss of sacB, and tetracycline. (c) Deletions resulting from selection for resistance to streptomycin, caused by loss of strA, and kanamycin. R/r indicates resistance and S/s indicates sensitivity. (Redrawn with permission from Krishnan *et al.* 1995.)

are now being generated at such a rate that a special database (dbEST) has been constructed specifically to meet the unique informatics challenges posed by cDNA sequences (ESTs). Two papers on ESTs from *Caernorhabditis* (Waterston *et al.* 1992; McCombie *et al.* 1992a) and one from rice (Kurata *et al.* 1994) show that the data generated are similar to that from humans.

Feature mapping

In feature mapping, large DNA fragments are cloned into a transposon-based cosmid vector designed for generating nested deletions by *in vivo* transposition and simple bacteriological selection. These deletions place primer sites throughout the DNA of interest at locations that are easily determined by plasmid size (Fig. 5.10). In this way, Krishnan *et al.* (1995) generated 70 informative deletions of a 35 kb human DNA fragment. DNA adjacent to the deletion end points was sequenced and constituted the foundation of a feature map by: (i) identifying putative exons and positions of *Alu* elements, (ii) determining the span of a gene sequence, and (iii) localizing evolutionary conserved sequences.

Alternative sequencing techniques

Because of the workload associated with generating long stretches (>100 kb) of sequence data, much of it technically demanding but extremely boring, attempts have been made to automate as much of it as possible. Automated sequencing machines have been available for a number of years but most of the labour is required in earlier steps, e.g. to select clones, make DNA preparations, carry out the sequencing chemistry and pour and load sequencing gels. Equipment that will automate some of these steps is beginning to become commercially available. However the Japanese, with their skills in instrumentation, have constructed an automatic DNA assembly line (Endo *et al.* 1991). Not only are all the manipulative steps undertaken by machine but samples are passed automatically from one machine to another. The cost to date of this set up is $10 million and its operational capacity has not been tested. Even if it is proved effective the price of this assembly line makes it unlikely it will be duplicated in many laboratories.

An alternative approach to increasing throughput is to develop new sequencing methodology. Over the years a number of novel techniques have been proposed. Despite generating a lot of initial excitement most have sunk without trace. Only two still are considered credible: scanning tunnelling microscopy (STM) and sequencing by hybridization. In STM, a very fine tip is kept extremely close to the object, in this case DNA, by a control system based on the detection of a minute current induced by tunnel effect between the tip and the DNA. In the related atomic force microscopy (AFM) control is based on measurement of the Van der Waals interaction force between the probe and the sample. Either way, the tip is moved to scan across the object, its vertical displacements being precisely measured and stored thereby generating a picture of a sample surface. Reasonable pictures have been taken of single-stranded DNA (Dunlapp & Bustamante 1989) and double-stranded DNA (Driscoll *et al.* 1990) but it remains to be seen if individual bases in DNA can be recognized. If they could then the technique has the potential to generate 1 Mb of sequence data in less than a day.

The continued refinement of two recent methods for producing gas-phase ions, electrospray ionization and matrix-assisted laser desorption ionization, has resulted in new techniques for the rapid characterization of oligonucleotides by mass spectrometry. However, it is considered unlikely that mass spectrometry will replace existing methods for sequencing oligos much longer than 100mers (Limbach *et al.* 1995).

Sequencing by hybridization

The principle of sequencing by hybridization can best be explained by starting with a simple example. Consider the tetranucleotide CTCA, whose complementary strand is TGAG, and a matrix of the whole set of $4^3 = 64$ trinucleotides. This tetranucleotide will specifically hybridize only with complementary trinucleotides TGA

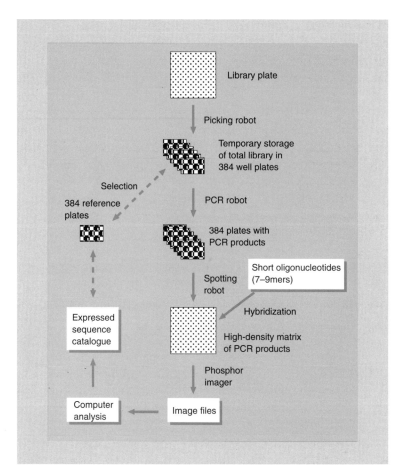

Figure 5.11 Outline of scheme for sequencing by hybridization where labelled probes are applied to cloned DNA immobilized on membranes. (Redrawn from Meier-Ewert *et al.* 1993, with permission from *Nature*, copyright (1993) Macmillan Magazines Ltd.)

and GAG revealing the presence of these blocks in the complementary sequence. From this can be reconstructed the sequence TGAG. If instead of using trinucleotides, $4^8 = 65\,536$ octanucleotides were used it should be possible to sequence DNA fragments up to 200 bases long (Bains & Smith 1988; Lysov *et al.* 1988; Southern 1988; Drmanac *et al.* 1989). Two different experimental configurations have been developed for the hybridization reaction. Either the target sequence may be immobilized and oligonucleotides labelled (Fig. 5.11) or the oligonucleotides may be immobilized and the target sequence labelled. Each method has advantages over the other for particular applications. It is an advantage to label the oligonucleotides to analyse a large number of target sequences for fingerprinting. On the other hand, for applications that require large numbers of oligonucleotides of different sequence, it is advantageous to immobilize oligonucleotides and use the target sequence as the labelled probe.

Drmanac *et al.* (1992) have developed a protocol based on the first method and which has the theoretical capability of determining up to 100 Mb per year. Indeed, 20 replica filters containing 0.5 million clones, representing 100 Mb, can be prepared by one individual in

103

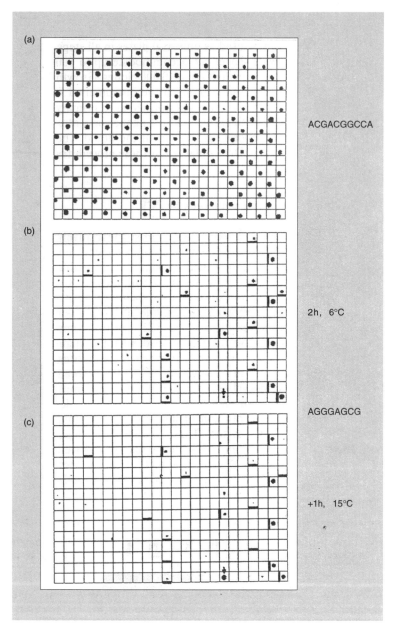

Figure 5.12 Discriminative hybridization of an eight-mer probe to M13 clones.
(a) 192 clones immobilized on a membrane and hybridized with a labelled vector
sequence to determine those spots carrying sufficient DNA for detection by
hybridization. (c) The same filter hybridized with the indicated eight-mer. The filter
is exposed after two successive washes. Dots marked by vertical bars contain fully
matched targets. Dots marked by horizontal bars contain G/T end-mismatched
duplexes which are discriminated more efficiently during the second wash. The
intensities of some dot-like spots which are not washed out increase relative to fully
matched signals. An example in the last row is marked by an arrow. After the first
wash, it is weaker than the full match and the end-mismatch in the same row. After
the second wash it is even stronger than the full match. (Reproduced with
permission from Drmanac *et al.* 1992.)

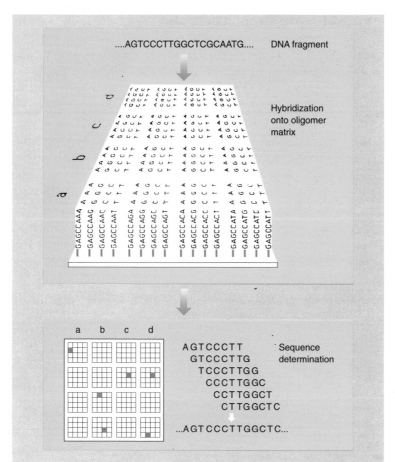

Figure 5.13 Schematic
representation of sequencing
by hybridization where the
sequence being examined is
labelled and hybridized with
immobilized oligomers.
(Redrawn with permission
from Hoheisel 1994.)

one month. Typical results that can be obtained are shown in Fig.
5.12. More recently, Drmanac *et al.* (1993) successfully used several
hundred 8–10mer probes to sequence three 343 base fragments of
DNA from phage M13. Meier-Ewert *et al.* (1993) have described
how this method can be used as an alternative to gel-based methods
to partially sequence cDNA clones (see p. 97). Southern *et al.* (1992)
have developed model systems based on the second configuration
(Fig. 5.13) and demonstrated its feasibility. They developed
technology for making complete sets of oligonucleotides of defined
length and covalently attaching them to the surface of a glass plate
by synthesizing them *in situ*. A device carrying all octapurine sequences was used to explore factors affecting molecular hybridization
of the tethered oligonucleotides, to develop computer-aided methods
for analysing the data, and to test the feasibility of using the method
for sequence analysis.

If large-scale sequencing of DNA by hybridization is to become a
reality then methods will have to be developed to conveniently and
accurately synthesize all 65 536 octanucleotides and lay these down

on a solid format. Good progress has been made by combining photolithographic techniques with solid-phase chemical synthesis. Pease *et al.* (1994) have synthesized miniature arrays of 256 eight-mers and demonstrated the availability of the probes on the surface for hybridization and the discrimination of perfect and imperfect complementarity. More important still, the synthesis of all 65 536 octanucleotides on a 1.28 × 1.28 cm microchip has been demonstrated (Huang, cited in Lipshutz & Fodor 1994). The success of these chip-based arrays will be dependent on the physical chemistry of the interaction between the octanucleotide probe and the target DNA sequence. Livshitz *et al.* (1992) have developed a model system for investigating such parameters as washing rate, melting temperature, probe concentration, etc. Although much development work still needs to be done on hybridization as a sequencing method it now appears realistic that it could become a viable alternative to gel-based methods.

6 Finding genes in large genomes

Introduction

Using classical genetics it is relatively easy to show the mode of inheritance of a particular Mendelian trait. Developing an understanding of the biological basis of the phenotype is harder for this requires a detailed knowledge of the appropriate gene(s), genetic control of the gene(s), identification of gene products and their role in the life of the host. In most instances this requires gene isolation using recombinant DNA technology. Where biochemical information is available it usually is possible to devise a suitable protocol for selecting the appropriate gene from gene libraries (see Old & Primrose 1994). This has been described as *functional cloning* (Collins 1992). However, for the vast majority of traits, including over 4000 in man, no biochemical information is available. Recently a number of tech-niques have been developed for *positional cloning* of genes of unknown function. Functional cloning depends upon application of biological information while positional cloning is initiated by mapping the respon-sible gene to its correct location on a chromosome (Fig. 6.1). Successive narrowing of the candidate interval eventually results in the identification of the correct gene whose function then can be studied.

Mapping the gene of interest

In practice, one starts with a collection of pedigrees in which the respon-sible gene is segregating. These families are studied with multiple poly-morphic markers until evidence for chromosomal location is obtained. Then linkage to other markers on that chromosome is determined. Clearly, the greater the number of markers available per chromosome, the more useful will be the mapping data. The closer the gene of interest to a known marker, the easier will be the later stages. In man, it is not uncommon for the separation of the two to be 1Mb (1cM) but in other organisms with much less repeated DNA this distance can be much shorter, e.g. in yeast 1cM is equivalent to 3 kb. Absolute location of the gene of interest is facilitated if the corresponding trait is associated with chromosome abnormalities that are cytogenetically visible, e.g. fragile sites, deletions, duplications and translocations.

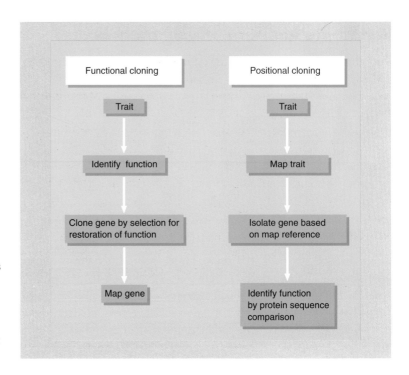

Figure 6.1 Comparison of
functional and positional
cloning. Functional cloning
depends on the availability
of information about the
protein product and/or
function of the responsible
gene. Mapping then follows
cloning. Positional cloning
moves in the opposite
direction. Here mapping is
essential to the process and
function is determined after
the gene is identified.

The next stage is chromosome walking (see p. 56) and jumping to
bridge the gap between the nearest markers and the gene of interest.
Although attractive in theory, chromosome walking suffers from the
problems that it is very laborious and long-range walking is not
practicable. In addition, the presence of unclonable DNA can bring
the walk to a grinding halt. Chromosome jumping avoids these
pitfalls. This involves circularization of very large genomic DNA
fragments followed by cloning DNA from the region covering the
closure site of these circles. This brings together DNA sequences that
were originally located a considerable distance apart in the genome.
These cloned DNAs from the closure sites make up a 'jumping
library'. Figure 6.2 shows a strategy (Poustka *et al.* 1987) for creating
a jumping library using *Not*I-digested human genomic DNA. This
enzyme cuts only rarely in mammalian DNA and so drastically
reduces the size of the library to cover the mammalian genome. This
makes analysis of jumps easy if PFGE and Southern blotting of *Not*I-
digested genomic DNA is used. However, with a complete digest
such as this, overlaps between neighbouring *Not*I fragments have to
be obtained indirectly. One solution is to use 'linking clones'
containing conventionally cloned fragments with an internal *Not*I
site. A variant of the strategy shown in Fig. 6.2 employs partial
digestion with a restriction endonuclease to generate large genomic
DNA fragments. This technique has been applied to a jump of 100 kb
in the cystic fibrosis locus (Collins *et al.* 1987) and to a jump of 200
kb in the region of the Huntington's disease gene (Richards *et al.*
1988).

Identifying the gene of interest

Once the locus of the gene of interest has been mapped, one or more YAC clones are isolated which contain sequences from the surrounding region and these are subcloned to generate a contig of overlapping phage or cosmid clones from which DNA can be easily purified. Several different approaches are available for gene identification (Table 6.1) and these are reviewed by Monaco (1994).

One approach is based on the observation that coding sequences are strongly conserved during evolution whereas non-coding DNA is not (see, for example, the *Drosophila* mapping experiment described on p. 60). A DNA clone that may contain a gene can be hybridized against a Southern blot of genomic DNA samples from a variety of species, a so-called zoo blot (Monaco *et al.* 1986). At reduced hybridization stringency probes containing human genes, for example, will generally show strong hybridization to animal

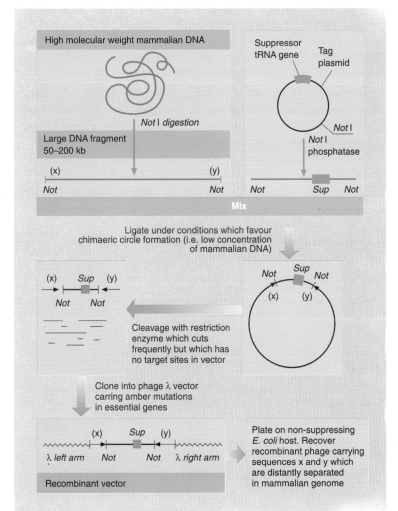

Figure 6.2 Construction of a jumping library using *Not*I-digested mammalian DNA. The suppressor tRNA gene acts as a selective marker for the recombinants because it suppresses the amber mutations in essential genes of the phage vector.

Table 6.1 Strategies to identify genes during positional cloning.	Traditional approaches	Newer approaches
	Conservation on zoo blots	Exon amplification
	CpG islands	cDNA selection
	Northern blots	Genomic sequencing and computer analysis
	cDNA library	Regionally mapped candidate genes

genomic DNA. Probes containing only human non-coding sequences will not. However, genes that are species-specific will not be detected in this way. By means of such zoo blots Sedlacek *et al.* (1993) isolated genomic DNA species conserved between humans, mice and pigs.

CpG islands are short stretches of DNA, often 1 kb or less, containing CpG dinucleotides which are unmethylated and which are present at the expected frequency, unlike in the remainder of the genome. In mammals such islands of normal CpG distribution are thought to mark transcriptionally active DNA sequences and are often found at the 5' end of housekeeping genes (Craig & Bickmore 1994). CpG islands can be recognized by diagnostic patterns of short- and long-range restriction maps. For example, *Hpa*II cleaves at the recognition site CCGG, but will not cleave CC^{M}GG. Thus *Hpa*II will tend to cut selectively at CpG islands hence these islands are often referred to as HTF islands (*Hpa*II tiny fragments). Rare cutter enzymes such as *Not*I (GCGGCCGC) often cleave within GpC island sequences because the islands are enriched in G+C and CpG dinucleotides. Clustering of such sites thus is commonly associated with CpG islands.

Conserved DNA fragments, CpG island-containing fragments and random single-copy DNA fragments can be tested for expression by hybridization to Northern blots of RNA isolated from fetal and adult tissues. If a transcript is identified then the genomic DNA fragment subsequently can be hybridized to cDNA libraries. Although Northern blotting and the analysis of conserved DNA fragments and CpG islands have been particularly effective in the isolation of many human disease genes (Table 6.2) they are very labour intensive. An alternative approach is to directly hybridize whole YAC or cosmid inserts to cDNA libraries on filters after suppression of repeated sequences in the radioactively labelled probe. Although technically difficult the method has enabled many new genes to be isolated (Monaco 1994).

One of the newer approaches to gene isolation is cDNA selection (Lovett *et al.* 1991; Parimoo *et al.* 1991). In this method, an amplified cDNA library is hybridized to immobilized YAC or cosmid clones that cover part of the genome of interest. Of those cDNAs isolated, at least one should correspond to the desired gene. Again, the process is technically difficult (Lovett 1994) because of simultaneous selection of cDNAs with homology to pseudogenes and low copy number repeated sequences. A variation of this method is to use exon

Disease locus	Cytogenetic rearrangement	
Chronic granulomatous disease	+	
Duchenne muscular dystrophy	+	
Retinoblastoma	+	
Wilms tumour	+	
Cystic fibrosis	–	
Neurofibromatosis type 1	+	
Testis determining factor	+	
Fragile X syndrome	+	
Familial polyposis coli	+	
Choroideremia	+	
Kallmann syndrome	+	
Aniridia	+	
Myotonic dystrophy	–	
X-linked agammaglobulinaemia	–	

Table 6.2 Some human disease genes identified by positional cloning.

amplification (Buckler *et al.* 1991). In this method, random segments of chromosomal DNA are inserted into an intron present within a mammalian expression vector. After transfection, cytoplasmic mRNA is screened by PCR amplification for the acquisition of an exon from the genomic fragment. The amplified exon is derived from the pairing of unrelated vector and genomic splicing signals. A number of human genes have been isolated successfully using this method (Monaco 1994).

As noted in the previous chapter (p. 97) a number of groups have begun large-scale sequencing of the human genome and analysis of assembled sequences for potential ORFs. If the gene of interest lies within a sequenced region then its identification should be relatively easy. The approach of sequencing genomic DNA to find expressed sequences has been applied to the positional cloning of the human gene for Kallmann syndrome. Whereas one group used conservation of sequences from YACs to identify potential exons (Franco *et al.* 1991), the other completely sequenced 60 kb of genomic DNA and identified potential coding regions (Legouis *et al.* 1991).

Isolation and characterization of the muscular dystrophy (DMD) gene as an example of positional cloning

The first successful application of positional cloning was the elucidation of the molecular basis of Duchenne muscular dystrophy (Hoffman *et al.* 1987). In this instance, isolation of the gene was facilitated by the availability of affected individuals with chromosomal translocations and deletions which localized the gene within the p21 region of the human X chromosome. The availability of a patient with a deletion was particularly fortuitous since it enabled subtraction cloning to be undertaken. For this, genomic DNA from a normal individual was digested with endonuclease *Mbo*I which produces fragments with

GATC overhangs. The *Mbo*I fragments were denatured and annealed with a 200-fold excess of denatured DNA from the deletion patient and which had been sonicated to produce ragged ends. Under these circumstances Xp21 DNA present in normal individuals but absent in the deletion patient will anneal with itself and will be the only DNA with a GATC overhang at each end. Thus it could be selectively cloned in a vector cut with *Bam*HI. Individual DNA clones then were used as probes in Southern blot hybridizations against normal and patient DNA samples to identify those which could be carrying the DMD gene. One of six fragments derived from the Xp21 region detected deletions in a proportion of patients with DMD and was tightly linked to the disease in family studies. Approximately 200 kb of contiguous DNA in this region was isolated by chromosome walking. The next step was to search for RNA transcripts of the suspected DMD gene. For this, non-repetitive DNA segments were screened by zoo blotting and one particular region, located 70 kb distant to the original clone, hybridized at high stringency with chicken as well as rodent and primate DNA. Sequencing of the homologous human and mouse segments disclosed an exon bounded by splicing signals. A probe against this region detected a large RNA species in human fetal muscle which was absent from other tissues. Antisera raised against expressed portions of the gene sequence cross-react with a 400 000 kDa protein, called dystrophin, present in normal adult and fetal muscle but which is absent in DMD patients.

The identification of dystrophin by positional cloning is a vindication of the concept that uncharacterized traits can be tackled by first finding the gene and working back to the product, i.e. genetics in reverse. It also shows how conventional genetics, recombinant DNA technology, sequencing and biochemical analysis all are required for success.

Chromosome landing

In positional cloning one finds a marker linked to a gene of interest and then walks to the gene via overlapping clones. As noted earlier (p. 57), chromosome walking is fraught with difficulties in organisms such as man. In plants with a more complex genome organization chromosome walking is not feasible. However, the strategy of chromosome walking is based on the assumption that it is difficult to find DNA markers that are physically close to a gene of interest. In plants this need not be the case, particularly since the advent of RAPDs (p. 68). The plant mapping paradigm now is the isolation of one or more DNA markers at a physical distance from the targeted gene that is less than the average insert size of the genomic library being used for clone isolation. The DNA marker then is used to screen the library and isolate, or land on, the clone containing the gene of interest (for review, see Tanksley *et al.* 1995).

A good example of chromosome landing is provided by the

isolation of the tomato gene *Pto* which confers resistance to the bacterial pathogen *Pseudomonas syringae* (Martin *et al.* 1993). Initially, 18 informative markers closely linked to the *Pto* gene were identified. High-resolution analysis in more than 1200 F_2 plants identified one marker that co-segregated with the gene. This marker was used to isolate YAC clones derived from the target region and genetic mapping of the isolated YAC ends identified one clone that spanned the *Pto* locus. This YAC clone was used to probe a cDNA library and all the cDNAs which hybridized were placed on a high-resolution linkage map. When this was done one group of cDNAs was identified which was closely linked to the *Pto* locus. When susceptible plants were transformed with this cDNA they acquired resistance to bacterial infection. This work exemplifies the advantages of chromosome landing in that the initial emphasis on the isolation of many closely linked DNA markers eliminated the need for chromo-some walking. The development of a high-resolution linkage map by conventional genetic analysis expedited the isolation of candidate cDNAs.

The positional candidate approach to gene identification

By the end of 1994 a large number of human genes had been isolated by positional cloning. Representative examples are given in Table 6.2 and a fuller listing is provided by Collins (1995). In many cases success depended on the presence of chromosomal deletions, translocations or trinucleotide repeat expansions. The latter are mutations caused by amplification (up to 54-fold) of a particular codon (Caskey *et al.* 1992). Only three genes, cystic fibrosis (Rommens *et al.* 1989), diastrophic dysplasia (Hastbacka *et al.* 1994) and *obese* (Zhang *et al.* 1994), have been positionally cloned based only on point mutations. Success in these cases was dependent on an extremely dense collection of markers close to the gene of interest coupled with fine-structure linkage disequilibrium mapping or, in the case of the mouse *obese* gene, directed genetic crossing. The success of Legouis *et al.* (1991) in identifying the Kallmann syndrome gene from DNA sequence data has led to an alternative approach designated the 'candidate gene' or 'positional candidate' approach (Ballabio 1993). Unlike functional and positional cloning this approach does not require the isolation of new genes but relies on the availability of information regarding function and map position from previously isolated genes (see Chapter 5). This information could have come from either genomic sequencing projects or expressed sequence tags/cDNA fingerprints. A good example is the human gene for X-linked glycerol kinase deficiency where two groups spent several years isolating it by positional cloning. Meanwhile, a third group (Sargent *et al.* 1993) mapped a randomly isolated EST, homologous to *Bacillus subtilis* glycerol kinase, and located it in the human Xp21 region. After obtaining a longer cDNA sequence it was shown to be the relevant gene. To date, 19 human disease genes have been identified by this method (Collins 1995). The candidate gene approach has been used in species other than man. For example, it was used to clone 31 out of 42 mutations in the mouse (Copeland *et al*. 1993) and to clone the gene for malignant hyperthermia in pigs (Fujii *et al*. 1991). An association also has been noted between a RFLP within the gene encoding the oestrogen receptor and litter size in pigs (Rothschild *et al*. cited by Archibald 1994a).

In future, when a new trait is assigned to a specific map position it may be possible to interrogate the genome database for that portion of the chromosome and obtain a list of other genes assigned to the same region. The features of the genes will then be compared with the features of the trait to find the most likely candidate gene. The potential gene from individuals exhibiting the trait can be screened by sequencing or hybridization to determine if it carries sequence abnormalities. This process can go in both directions: from trait to genes and from genes to traits.

Studying complex and quantitative traits

The term 'complex trait' refers to any phenotype that does not exhibit classic Mendelian recessive or dominant inheritance attributable to a single gene locus. Most, but not all, complex traits can be explained by polygenic inheritance. That is, these traits require the simultaneous presence of mutations in multiple genes. Polygenic traits may be classified as discrete traits, measured by a specific outcome (e.g. development of diabetes or death from myocardial infarction), or quantitative traits measured by a continuous variable (Lander & Schork, 1994).

Many important biological characteristics are inherited quantitatively but because these effects have not generally been resolvable individually, quantitative geneticists have dealt largely with characterization of these factors using biometrical procedures. However, many issues in quantitative genetics and evolution are difficult to address without additional information about the genes that underlie continuous variation. The identification of such *quantitative trait loci* (QTLs) has become possible with the advent of restriction fragment length polymorphisms (RFLPs) as genetic markers and the increasing availability of complete RFLP maps in many organisms.

There are two principal approaches to mapping QTLs. Both involve arranging a cross between two inbred strains differing substantially in a quantitative trait. Segregating progeny are scored both for the trait and for a number of genetic markers. Typically the segregating progeny are produced by a B_1 backcross ($F_1 \times$ parent) or an F_2 intercross ($F_1 \times F_1$). In the method of Edwards *et al*. (1987), linear regression is used to examine the relationship between the performance for the quantitative trait and the genotypes at the marker locus. If there is a statistically significant association between the trait performance and the marker locus gene types (Fig. 6.3) it is inferred that a QTL is

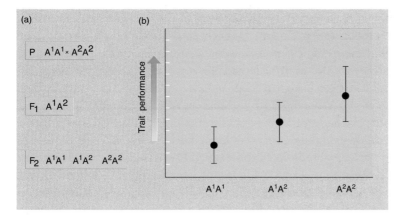

Figure 6.3 Relationship between the performance for a quantitative trait and the genotypes at marker locus for an F_2 interbreeding population. (a) Genetic composition of the F_2 population sampled. (b) Quantitative performance of different types of F_2 genotypes.

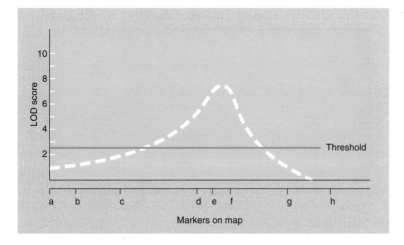

Figure 6.4 Interval mapping to determine if QTL is located near a particular genetic marker. The distances between the genetic markers represent genetic distance.

located near the marker locus. Lander and Botstein (1989) have pointed out that there are a number of disadvantages with this method. In particular, it cannot distinguish between tight linkage to a QTL with small effect and loose linkage to a QTL with large effect. This is not a problem with the second method, called *interval mapping*, which is an adaptation of the LOD score analysis used in human genetics. This involves calculating the ratio of the likelihood (odds) that there is a QTL to the likelihood that there is no QTL at each position along the length of the chromosome, e.g. from marker loci a to e in Fig. 6.4. These values normally are given as the $\log_{10}$ of the 'odds' ratio or LOD score. On the basis of the size of the genome and the number of marker loci analysed, a threshold value or significance level is determined. In the example shown in Fig. 6.4 this threshold value has been set at 2.5. Where the LOD score exceeds the threshold a QTL is likely. Recently, Haley *et al.* (1994) have adapted the interval mapping approach for mapping QTLs in outbreeding populations.

The above methods have been used to map QTLs in both crops and livestock animals. For example, Paterson *et al.* (1988) mapped 15 QTLs in the tomato affecting fruit mass, concentration of soluble solids and fruit pH. Andersson *et al.* (1994) generated an F_2 intercross between wild boars and domestic pigs. Two hundred offspring were genotyped for 105 polymorphic markers. In addition, growth rates were recorded from birth until they weighed 70 kg, and after slaughter fat levels and intestinal length were noted. Fatness, whether measured as a percentage of fat in the abdominal cavity or backfat thickness, mapped to the proximal end of chromosome 4. The QTL for intestinal length and growth rate were located distal to the fatness QTL.

In the pig study noted above, the QTLs have been mapped to regions of the genome 15.25 cM in length. Thus the use of positional cloning for gene identification is not feasible (see p. 113). The positional candidate-gene approach is much more likely to be successful (Archibald 1994a). However, this approach depends on the linkage map being well populated with known genes. In the absence of these, the comparative gene maps described in Chapter 1 (p. 6) assume far greater significance.

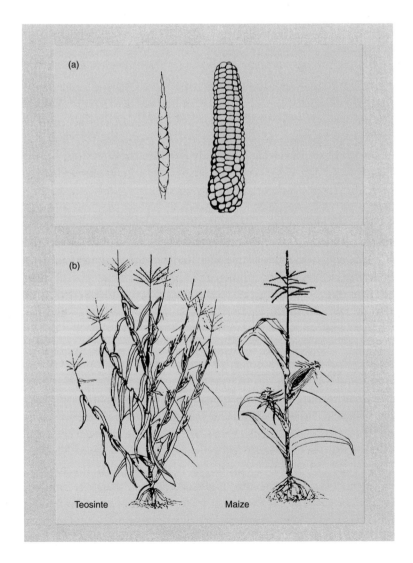

Figure 6.5 Comparison of (a) ear structure, and (b) plant morphology for maize and its ancestor teosinte. (Adapted and redrawn with permission from Doebley 1992.)

The analysis of complex traits in humans demands a different approach. One has to start with the analysis of families of affected individuals. Affected members of a family will share the genes causing the disease and this sharing will manifest itself as an excess of sharing over the 50% expected at random in siblings. Disease genes can be identified by co-segregation with linked genetic markers which will show identity by descent in affected individuals. The markers used must be close enough that recombination with the disease locus is unlikely and sharing among individuals can only be tested if the markers are polymorphic. Microsatellite markers are ideal for this purpose as they are highly polymorphic and sufficient numbers of them have been positioned throughout the genome. Using such microsatellite markers, Davies *et al.* (1994) and Hashimoto *et al.* (1994) have carried out a genome-wide search for genes conferring susceptibility to human type 1 (insulin dependent) diabetes. Both groups confirmed the role of IDDM1, a gene of the major histo-compatibility locus, and IDDM2, believed to be the insulin locus itself. They also identified a new susceptibility locus in the vicinity of the FGF3 gene on the long arm of chromosome 11. More recently, Copeman *et al.* (1995) have used linkage disequilibrium in combination with microsatellite markers to map IDDM7 to chromosome 2.

The development of methods for mapping and characterizing QTLs can be used to study evolution at the molecular level. A good example is provided by maize (Indian corn) and its relatives, the teosintes. Although they differ in both ear (corn cob) morphology and plant growth form (Fig. 6.5) it is believed that maize is a domestic form of teosinte. Indeed, a small number of genes selected by the pre-historic peoples of Mexico may have transformed teosinte into maize within the past 10 000 years. Morphologically there are five points of difference between the two plants and analysis by classical genetics has suggested that five major and independently inherited gene differences distinguish the two species. Doebley (1992) has confirmed this by molecular methods and has shown that the major differences map to five restricted regions of the genome, each on a different chromosome.

Genome scanning methods

Positional cloning is a powerful approach to the isolation of genes which is applicable, in principle, to any organism. In practice, its actual use has been much more restricted because success is dependent on the ability to find tightly linked genetic markers near a locus of interest. Thus the method is restricted to those organisms for which dense genetic maps have been constructed. Even then, determining linkage can be a lot of work. Other methods of gene isolation depend on some knowledge of the function of the gene product and, again, this may not be available. Ideally, what is required are methods that

117

allow an entire genome to be scanned at one time for sequences linked to trait loci. A number of these, termed *genome scanning methods* (Brown 1994), have been developed and are described below. Interestingly, with them, classical genetic techniques are as important as the molecular methods used.

GENETICALLY DIRECTED REPRESENTATIONAL DIFFERENCE ANALYSIS (GDRDA)

GDRDA is based on a subtractive technique, known as represen-

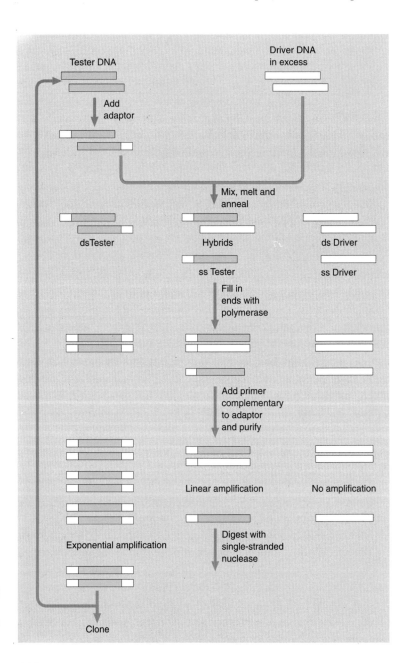

Figure 6.6 The principle of representational difference analysis (RDA). See text for details.

tational difference analysis (RDA), for identifying differences between two DNA samples (Lisitsyn *et al.* 1993). The two DNA samples are known as Tester and Driver. Specifically, RDA is designed to clone restriction fragments that can be amplified by PCR from Tester but not Driver. This may be because the corresponding sequence is completely absent from the Driver due to a homozygous deletion or because it is contained in a small restriction fragment in the Tester but in a large and, therefore poorly amplifiable, restriction fragment in the Driver. Thus RDA exploits size differences in DNA from Driver and Tester.

The principle of RDA is shown in Fig. 6.6. Restriction-enzyme-digested genomic DNA fragments are first ligated to adaptor oligonucleotides to allow their amplification using a cognate primer in PCR. Because of the biased amplification of fragments in the initial complex pool principally favouring fragments of smaller size, the pool, of products (an 'amplicon') obtained after PCR amplification of the initial sample constitutes a 'representation' of the genome. The complexity of this may be substantially less than that of the original genomic DNA. Ligation of a second adaptor oligonucleotide to the 3' ends of the fragments in one of the representations (the 'tester' amplicon) provides a means for selectively amplifying duplex DNA molecules both strands of which derive from the 'tester' amplicon. Thus, after denaturation and hybridization in the presence of an excess of a second amplicon ('driver'), sequences represented in the tester, but not the driver, amplicon are preferentially recovered in a form that can be amplified by subsequent PCR. Successive rounds of selection allow further subtractive hybridization and exploit the amplification of the intial enrichment factor that results from the dependence of the annealing rate of each fragment on its relative abundance in the pool (kinetic enrichment).

Recently, Rosenberg *et al.* (1994) described another method, RFLP subtraction, designed to purify restriction fragments from a complex genome if they do not have a counterpart of the same size in a competing genome. RFLP subtraction uses gel purification rather than PCR for size fractionation of DNA fragments. Rosenberg *et al.* (1994) took a slightly different approach for selective amplification of fragments unique to the target DNA. Before PCR amplification they performed three repetitive cycles of annealing the target with biotinylated driver and depleting biotinylated sequences with avidin binding. Once fragments unique to the target had been concentrated sufficiently, the method involved was identical to RDA. These added steps appeared to increase the power and specificity of the procedure: 21 of 22 clones of RFLP subtraction products represented unique RFLPs.

It is not enough to apply RDA to samples from a single affected and a single unaffected individual in a population or family. The abundant genetic variation among even close relatives means that polymorphisms will be found throughout the genome. A method of finding polymorphisms specifically in the vicinity of the gene of interest is required and this is provided by classical genetic crosses. Suppose A and B are two inbred strains differing at a target locus L, strain A carrying a mutant recessive allele and B carrying a dominant wild type allele. To create a Driver sample, an F_2 intercross between the strains is performed, a collection of progeny showing the recessive phenotype is selected, and their DNA mixed together. From Mendelian genetics it can be predicted that the Driver contains (i) no B alleles in the immediate vicinity of L, because progeny were selected for the recessive phenotype; (ii) a deficit of B alleles in a somewhat larger region around L, owing to linkage to L; and (iii) roughly equal proportions of A and B alleles elsewhere in the genome. If RDA is performed with this Driver and DNA from strain B as Tester then B alleles should be subtracted everywhere in the genome except in a region around L.

The targeting of the method can be improved further if locus L has been genetically mapped and two flanking markers (X and Y) identified. For the Driver k/2 progeny are selected in which cross-over occurred between L and X and k/2 progeny in which the cross-over occurred between L and Y (where k is the total number of progeny). This refinement of the method should allow targeting of very small intervals of DNA and Lisitsyn *et al.* (1994) confirmed this by generating probes mapping close to two target mouse loci. GDRDA is applicable to more than just F_2 inter-crosses between inbred strains. It can be applied to backcrosses between inbred strains, two-generation families in an outbred population (e.g. humans where inbred lines are not readily available) and half-sib mating schemes (common in livestock breeding).

To date, the use of RDA and RFLP subtraction in plant studies has not been reported. However, in the technique called chromosome landing (p. 112), random amplified polymorphic DNAs (RAPD) have been used in a similar way (for review see Tanksley *et al.* 1995).

GENOME MISMATCH SCANNING (GMS)

Whereas GDRDA focuses on dissimilarities between two genomes, GMS seeks to identify large regions of sequence identity between two individuals (Nelson *et al.* 1993). The rationale is that because the rate of natural polymorphism in a population is substantial identical sequences must represent genomic regions of 'identity by descent'. The principle of GMS is shown in Fig. 6.7. High molecular weight DNA is prepared from two related individuals. Both genomic DNAs are digested with the same restriction endonuclease which must leave protruding 3' ends. The 3' protrusions serve to protect these ends from digestion with exonuclease III in later steps. Also, the DNA fragments need to be long enough so that many fragments

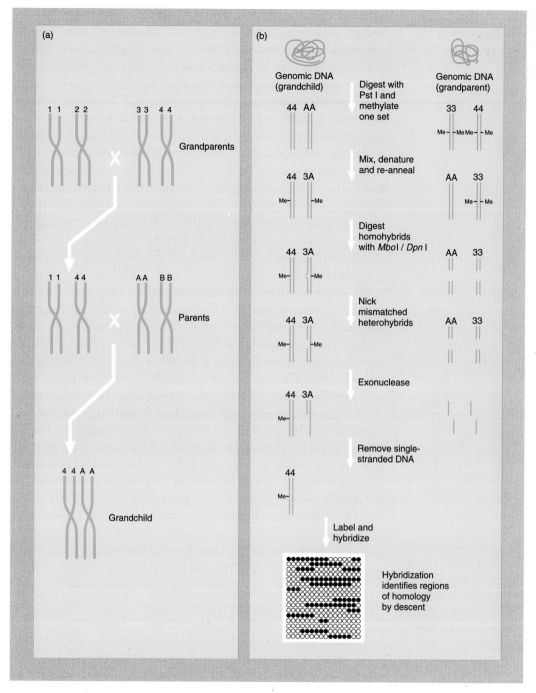

Figure 6.7 The principle of genome mismatch scanning (GMS). (a) The relationship between the chromosomes of the grandchild and grandparents. (b) Experimental method to generate regions of homology. See text for details.

are likely to contain both a GATC sequence and at least one base that differs between unrelated allelic fragments. To differentiate the two DNAs, one individual's DNA is fully methylated at all GATC sites with the *E. coli* Dam methylase. The other individual's DNA remains unmethylated at all GATC sites. The two restriction-digested genomic DNA samples are then mixed in equimolar ratios, denatured and allowed to reanneal. Approximately 50% of the reannealed fragments are heterohybrids and are, therefore, hemimethylated (methylated on only one strand) at each GATC sequence. 'Homohybrids', formed from two strands of the same individual's DNA, are either methylated on both strands or not methylated at all. Digestion of the reannealed DNA with both *Dpn*I and *Mbo*I, which cut at fully methylated and unmethylated GATC sites respectively, results in cleavage of the homohybrids to yield smaller duplexes with either blunt ends or 5' protruding ends. The heterohybrids are resistant to cleavage by both *Dpn*I and *Mbo*I. This step therefore distinguishes between homohybrids and heterohybrids.

Sensitive distinction between mismatch-free heterohybrids and those with base mismatches is achieved using the methyl-directed mismatch repair system of *E. coli*. Three proteins, MutL, MutS and MutH, are sufficient to recognize seven of the possible eight single-base-pair mismatches in duplex DNA. They are able to introduce a single strand nick on the unmethylated strand at GATC sites specifically in the mismatch-containing duplexes. The heterohybrids are incubated with MutL, MutS and MutH. Only those that are mismatch-free, and therefore likely to be from regions of 'identity by descent', escape nicking.

All DNA molecules, except the mismatch-free heterohybrids, can be degraded with ExoIII, a 3' to 5' exonuclease specific for double-stranded DNA (dsDNA). ExoIII can initiate digestion at a nick and degrade one strand to yield a single-stranded DNA (ssDNA) gap. ExoIII can also initiate at the blunt or 5' protruding ends produced by cleavage with *Dpn*I or *Mbo*I, respectively, to degrade the homohybrids to ssDNA. The heterohybrids have only 3' protruding ends which resist degradation by exonuclease III. Thus, only DNA duplexes which are both heterohybrids and free of mismatches escape digestion. Molecules containing single-stranded DNA are removed by chromatography. The purified heterohybrids are labelled and used to probe an ordered array of DNAs representing the entire genome. Regions of hybridization represent regions of identity by descent.

Nelson *et al.* (1993) have tested GMS experimentally using *Saccharomyces cerevisiae* as a model system. In their test, two parental yeast strains, A and B, were mated and the resulting diploid was sporulated to produce haploid progeny. Progeny were compared to each parent using GMS, with the resulting DNA being hybridized to a collection of consecutive phage clones covering the length of chromosome 5. As expected, the chromosome can be divided into blocks that are identical to either the A or the B parent, with the transitions corresponding to the site of cross-over (Fig. 6.8). The results were unambiguous for 20 of 31 phage clones. The remaining 11 clones were uninformative for this GMS-based linkage analysis because they gave consistently

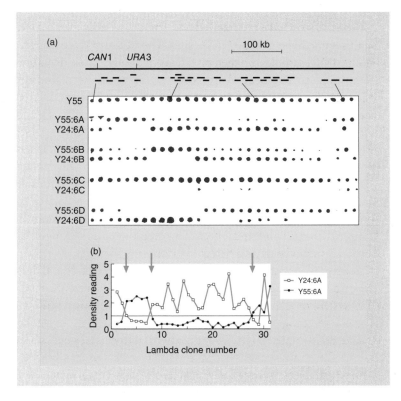

Figure 6.8 Example of the use of GMS in *Saccharomyces cerevisae*. GMS selected DNA from parent Y55 and Y24/spore clone (6A, B, C and D) pairs were hybridized onto an ordered array of 31 λ clones representing 79% coverage of chromosome 5. (a) The intervals spanned by each λ clone are indicated by bars above the dot blot. *CAN1* and *URA3* are two yeast genes mapped to chromosome 5. Row 1, Y55, shows the result of hybridization using unselected DNA from the Y55 parent as a probe to control for variability in the amount of DNA immobilized at each dot. Rows 2–9 show results of hybridization using GMS-selected DNA from the indicated parent:spore clone pair. (b) Normalized densitometer reading of the hybridization signals from the Y24:6A and Y55:6A pairwise comparisons. Meiotic recombination breakpoints are indicated by the arrows. (Redrawn with permission from Nelson *et al.* 1993 courtesy of *Nature Genetics*.)

positive or consistently negative results, most likely because of technical artefacts. If GMS can be applied satisfactorily to much larger genomes then it could facilitate the study of polygenic traits. A search would be made for regions of 'identity by descent' in pairs of affected cousins or in a collection of affected individuals from an isolated subpopulation.

SCREENING BY ENZYME MISMATCH CLEAVAGE (EMC)

The two genome scanning methods described above are not ideal. Although RDA has been shown to identify a specific locus in highly inbred mouse strains it really is not applicable to highly outbred populations such as man. GMS is effective in yeast but has not yet been shown to work on mammalian genomes which are much more complex and contain repetitive sequences. In addition, the unique sequences in the human genome are less variable between individuals than yeast sequences. By contrast, EMC has been developed for identifying mutations in human genes (Mashal *et al.* 1995; Youil *et al.* 1995).

Bacteriophage resolvases are enzymes whose function *in vivo* is to cleave branched DNA. They have the property of recognizing mismatched bases in double-stranded DNA and cutting the DNA at the mismatch. EMC takes advantage of this characteristic to detect individuals who are heterozygous at a given site. Radiolabelled DNA

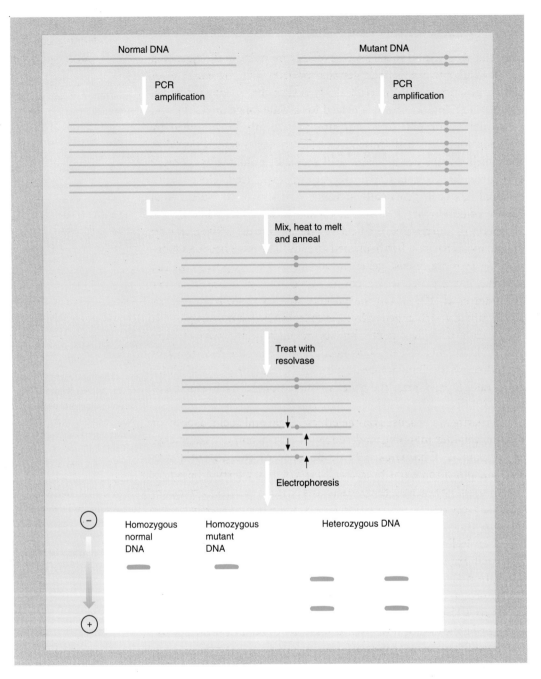

Figure 6.9 The principle of enzyme mismatch cleavage. See text for details.

is cleaved by the resolvase at the site of mismatch in hetero-duplex DNA and digestion is monitored on a gel (Fig. 6.9). Thus, both the presence and the estimated position of an alteration are revealed. In an analysis of all possible mismatch types as well as several deletion mutations, an overall sensitivity of mutation detection of 94% was achieved.

124

Ethical, legal and social issues

The ethical, legal and social aspects of gene mapping and sequencing are particular to the human genome project. The current state of clinical genetics (see Weatherall 1991) is such that pregnant women and newborn infants can be screened for a wide range of genetic and congenital disorders. This will almost certainly be followed by screening for genetic risk (prognostics). The range of genetic disorders for which screens are available will increase dramatically as more and more disease genes are mapped and sequenced. Once an individual has been genetically screened and shown to have a high chance of developing a particular disease in later life, what are the prospects of purchasing medical or life insurance? This is not an imaginary scenario, it has happened already. The issue becomes more pressing as more disease genes are identified and mapped, e.g. The *BRCA1* and 2 genes which confer susceptibility to breast cancer (Wooster *et al.* 1994). On the positive side, as more disease genes are identified and their gene products characterized, it may be possible to generate new therapeutic agents or appropriate gene therapy protocols. In this respect, it is encouraging that the first trials of gene therapy to eliminate cystic fibrosis (Crystal *et al.* 1994) have come only 5 years after the gene was first isolated by Rommens *et al.* (1989).

International consensus on the ethical, legal and social aspects of human genome mapping and sequencing is beginning to emerge (Knoppers & Chadwick 1994). Because of the potential for discrimination there has been a call for voluntary testing based on autonomous choice with the participants having the full information. Because most genetic information is only predictive and probabilistic it is important to protect the individual against social pressures. There is consensus limiting genetic testing, including prenatal screening, to tests that are medically therapeutic and France and Norway have legislation that defines such 'therapeutic' criteria. There also is consensus on two other issues. Predisposition testing should be limited to diseases that are treatable or preventable and presymptomatic testing of children for late onset disease is not recommended in the absence of treatment or prevention.

Respect for the privacy of the person and for the confidentiality of genetic information is crucial. Results of genetic tests differ significantly from other pieces of medical information in that they reveal information about other family members and this is of importance to insurers and employers. While there is consensus about not releasing these data to non-family members without consent there is a faction that advocates communication to family members at high risk without the consent of the patient. While there is consensus on restricting access to information by insurers, only Belgium has embodied this in law.

It is important that the scientific and medical community realize that much of the genetic analysis which is undertaken is technically complex. This means that the diagnostic use of genetic analysis should only be undertaken in accredited laboratories with properly trained staff who are aware of the sensitivity and specificity of the methods employed. They need to be monitored routinely and should operate within ethical guidelines which are clearly laid down.

References

Abidi F., Wada M., Little R.D. and Schlessinger D. (1990) Yeast artificial chromosomes containing human Xq24–Xq28 DNA: library construction and representation of probe sequences. *Genomics* 7, 363–376.

Adams M.D. and 12 others (1991) complementary DNA sequencing: expressed sequence tags and human genome project. *Science* 252, 1651–1656.

Adams MD., Dubnick M., Kerlavage A.R., Morenoo R., Kelley J.M., Utterback T.R. Nagle J.W., Fields C. and Venter J.C. (1992) Sequence identification of 2,375 human brain genes. *Nature* 355, 632–634.

Adams M.D., Kerlavage A.R., Fields C. and Venter J.C. (1993a) 3,400 new expressed sequence tags identify diversity of transcripts in human brain. *Nature Genetics* 4, 256–267.

Adams M.D., Soares M.B., Kerlavage A.R., Fields C. and Venter J.C. (1993b) Rapid cDNA sequencing (expressed sequence tags) from a directionally cloned human infant brain cDNA library. *Nature Genetics* 4, 373–380.

Alford R.L. and Caskey C.T. (1994) DNA analysis in forensics, disease and animal/plant identification. *Current Opinion in Biotechnology* 5, 29–33.

Altschul S.F., Boguski M.S., Gish W. and Woolton J.C. (1994) Issues in searching molecular databases. *Nature Genetics* 6, 119–129.

Anand R., Riley J.H., Butler R., Smith J.C. and Markham A.F. (1990) A 3.5 genome equivalent multi-access YAC library: construction, characterisation, screening and storage. *Nucleic Acids Research* 18, 1951–1956.

Andersson L. and 11 others (1994) Genetic mapping of quantitative trait loci for growth and fatness in pigs. *Science* 263, 1771–1774.

Archibald A.L. (1994a) Fat pigs can blame their genes. *Current Biology* 4, 728–730.

Archibald A.L. (1994b) Mapping of the pig genome. *Current Opinion in Genetics and Development* 4, 395–400.

Baer R. and 11 others (1984) DNA sequence and expression of the B95.8 Epstein–Barr virus genome. *Nature* 310, 207–211.

Bains W. and Smith G.C. (1988) A novel method for nucleic acid sequence determination. *Journal of Theoretical Biology* 135, 303–307.

Ballabio A. (1993) The rise and fall of positional cloning. *Nature Genetics* 3, 277–279.

Bankier A.T. and Barrell B.G. (1983) Shotgun DNA sequencing. *Technol. Nucleic Acid Biochemistry* B5, 1–34.

Barbour A.G. (1983) Linear DNA of *Borrelia* species and antigenic variation. *Trends in Microbiology* 1, 236–239.

References

Barendse W. and 22 others (1994) A genetic linkage map of the bovine genome. *Nature Genetics* **6**, 277–235.

Barrell B. (1991) DNA sequencing: present limitations and prospects for the future. *FASEB Journal* **5**, 40–45.

Barrett J.H. (1992) Genetic mapping based on radiation hybrid data. *Genomics* **13**, 95–103.

Belfort M. (1989) Bacteriophage introns: parasites within parasites? *Trends in Genetics* **5**, 209–213.

Bellanné–Chantelot C. and 21 others (1992) Mapping the whole human genome by fingerprinting yeast artificial chromosomes. *Cell* **70**, 1059–1068.

Bender W., Spierer P. and Hogness D.S. (1983) Chromosome walking and jumping to isolate DNA from the *Ace* and *rosy* loci and the bithorax complex in *Drosophila melanogaster*. *Journal of Molecular Biology* **168**, 17–33.

Benson D.A., Boguski M., Lipman D.J. and Ostell J. (1994) GenBank. *Nucleic Acids Research* **22**, 3441–3444.

Bird A. (1995) Gene number, noise reduction and biological complexity. *Trends in Genetics* **11**, 94–99.

Boguski M.S., Lowe T.M.J. and Tolstoshev C.M. (1993) dbEST–database for 'expressed sequence tags'. *Nature Genetics* **4**, 332–333.

Botstein, D., White R.L., Skolnick M. and Davis R.W. (1980) Construction of a genetic linkage map in man using restriction fragment length polymorphisms. *American Journal of Human Genetics* **32**, 314–331.

Branscomb E., Slezak T., Pae E., Gales D., Carrano A.V. and Waterman M. (1990) Optimizing restriction fragment fingerprinting methods for ordering large genomic libraries. *Genomics* **8**, 351–366.

Breatnach R., Mandel J.L. and Chambon P. (1977) Ovalbumin gene is split in chicken DNA. *Nature* **270**, 314–319.

Britten R.J. and Kohne D.E. (1968) Repeated sequences in DNA. *Science* **161**, 529–540.

Brown P.O. (1994) Genome scanning methods. *Current Opinion in Genetics and Development* **4**, 366–373.

Brown W.R.A. (1992) Mammalian artificial chromosomes. *Current Opinion in Genetics and Development* **2**, 479–486.

Buckler A.J., Chang D.D., Graw S.L., Brook D., Haber D.A., Sharp P.A. and Houseman D.E. (1991) Exon amplification: a strategy to isolate mammalian genes based on RNA splicing. *Proceedings of the National Academy of Sciences, USA* **88**, 4005–4009.

Buetow K.H., Weber J.L., Ludwigsen S., Scherpbier-Heddema T., Duyk G.M., Sheffield V.C., Wang Z. and Murray J.C. (1994) Integrated human genome-wide maps constructed using the CEPH reference panel. *Nature Genetics* **6**, 391–393.

Burke D.T., Carle G.F. and Olson M.V. (1987) Cloning of large segments of exogenous DNA into yeast by means of artificial chromosome vectors. *Science* **236**, 806–813.

Burr B. and Burr F.A. (1991) Recombinant inbreds for molecular mapping in maize: theoretical and practical considerations. *Trends in Genetics* **7**, 55–60.

Cantor C.R., Smith C.L. and Mathew M.K. (1988) Pulsed–field gel electrophoresis of very large DNA molecules. *Annual Review of Biophysical*

Chemistry 17, 287–304.

Caskey C.T., Pizzuti A., Fu Y-H., Fenwick R.G. and Nelson D.L. (1992) Triplet repeat mutations in human disease. *Science* 256, 784–789.

Chang C., Bowman J.L., De John A.W., Lander E.S. and Meyerowitz E.M. (1988) Restriction fragment length polymorphism linkage map for *Arabidopsis thaliana*. *Proceedings of the National Academy of Sciences, USA* 85, 6856–6860.

Chu G., Vollrath D. and Davis R. (1986) Separation of large DNA molecules by contour clamped homogenous electric fields. *Science* 234, 1582–1585.

Chumakov I.M. and 14 others (1992a) Isolation of chromosome 21–specific yeast artificial chromosomes from a total human genome library. *Nature Genetics* 1, 222–225.

Chumakov I. and 35 others (1992b) Continuum of overlapping clones spanning the entire human chromosome 21q. *Nature* 359, 380–387.

Church G.M. and Kieffer-Higgins S. (1988) Multiplex DNA sequencing. *Science* 240, 185–188.

Clarke L. and Carbon J. (1976) A colony bank containing synthetic Col E1 hybrid plasmids representative of the entire *E. coli* genome. *Cell* 9, 91–99.

Coe E., Hoisington D. and Chao S. (1990) Gene list and working maps. *Maize Genetics Cooperative Newsletter* 64, 154–163.

Cole S.T. and Saint Girons I. (1994) Bacterial genomes. *FEMS Microbiology Reviews* 14, 139–160.

Coletta P.L., Lench N.J., Brown K.A. and Markham A.F. (1994) Comparative genetic mapping for the identification of novel diagnostic and therapeutic targets. *Current Opinion in Biotechnology* 5, 643–647.

Collins F. and Galas D. (1993) A new five-year plan for the U.S. human genome project. *Science* 262, 43–46.

Collins F.S. (1992) Positional cloning: let's not call it reverse any more. *Nature Genetics* 1, 3–6.

Collins F.S. (1995) Positional cloning moves from perditional to traditional. *Nature Genetics* 9, 347–350.

Collins F.S., Drumm M.L., Cole J.L., Lockwood W.K., Van de Wonde G.F. and Iannuzzi M.C. (1987) Construction of a general human chromosome jumping library, with application to cystic fibrosis. *Science* 235, 1046–1049.

Copeland N.G. and 12 others (1993) A genetic linkage map of the mouse: current applications and future prospects. *Science* 262, 57–66.

Copeman J.B. and 15 others (1995) Linkage disequilibrium mapping of a type 1 diabetes susceptibility gene (IDDM7) to chromosome 2q31–q33. *Nature Genetics* 9, 80–85.

Coulson A. (1994) High-performance searching of biosequence databases. *Trends in Biotechnology* 12, 76–80.

Coulson A., Sulston J., Brenner S. and Karn L. (1986) Toward a physical map of the genome of the nematode *Caenorhabditis elegans*. *Proceedings of the National Academy of Sciences, USA* 83, 7821–7825.

Cox D.R., Burmeister M., Price E.R., Kim S. and Myers R.M. (1990) Radiation hybrid mapping: a somatic cell genetic method for constructing high–resolution maps of mammalian chromosomes. *Science* 250, 245–250.

Cox D.R., Green E.D., Lander E.S., Cohen D. and Myers R.M. (1994)

Assessing mapping progress in the human genome project. *Science* **265**, 2031–2032.

Craig J.M. and Bickmore W.A. (1994) The distribution of CpG islands in mammalian chromosomes. *Nature Genetics* **7**, 376–381.

Crystal R.G., McElvaney N.G., Rosenfeld M.A., Chu C–S., Mastrangeli A., Hay J.G., Brody S.L., Jaffe H.A., Eissa N.T. and Danel C. (1994) Administration of an adenovirus containing the human *CFTR* cDNA to the respiratory tract of individuals with cystic fibrosis. *Nature Genetics* **8**, 42–51.

D'Eustachio P. and Ruddle F.H. (1983) Somatic cell genetics and gene families. *Science* **220**, 919–928

Dausset J., Cann H., Cohen D., Lathrop M., Lalouel J.M. and White R.L. (1990) Collaborative genetic mapping of the Human Genome. *Genomics* **6**, 575–577.

Davidson E.H. and Britten R.J. (1973) Organization, transcription and regulation in the animal genome. *Quarterly Review of Biology* **48**, 565–613.

Davies J.L and 15 others (1994) A genome-wide search for human type 1 diabetes susceptibility genes. *Nature* **371**, 130–136.

Davies K.E., Young B.D., Elles R.G., Hill M.E. and Williamson R. (1981) Cloning a representative genomic library of the human X chromosome after sorting by flow cytometry. *Nature* **293**, 374–376.

Diehl S.R., Ziegle J., Buck G.A., Reynolds T.R. and Weber J.L. (1990) Automated genotyping of human DNA polymorphisms. *American Journal of Human Genetics* **47**, A177.

Dietrich W.F. and 14 others (1994) A genetic map of the mouse with 4006 simple sequence length polymorphisms. *Nature Genetics* **7**, 220–225.

Dimmock N.J. and Primrose S.B. (1994) *An Introduction to Modern Virology.* Blackwell Scientific Publications, Oxford.

Doebley J. (1992) Mapping the genes that made maize. *Trends in Genetics* **8**, 302–307.

Dogget N. (1992) In *The Human Genome Project.* Los Alamos Science No. 20. University Science Books, Sausalito, CA. 338 pp.

Donis-Keller H. and 32 others (1987) A genetic linkage map of the human genome. *Cell* **51**, 319–337.

Driscoll R.J., Youngquist M.G. and Baldeschweiler J.D. (1990) Atomic-scale imaging of DNA using scanning tunnelling microscopy. *Nature* **346**, 294–296.

Drmanac R., Drmanac S., Labat I., Crkvenjakov R., Vincentic A. and Gemmell A. (1992) Sequencing by hybridization: towards an automated sequencing of one million M13 clones arrayed on membranes. *Electrophoresis* **13**, 566–573.

Drmanac R., Drmanac S., Strzoska Z., Paunesku T., Labat I., Zeremski M., Snoddy J., Funkhouser W.K., Koop B., Hood L. and Crkvenjakov R. (1993) DNA sequencing determination by hybridization: a strategy for efficient large scale sequencing. *Science* **260**, 1649–1652.

Drmanac R., Labat I., Brukner I. and Crkvenjakov R. (1989) Sequencing of megabase plus DNA by hybridization: theory of the method. *Genomics* **4**, 114–128.

Dujon B. and 107 others (1994) Complete DNA sequence of yeast chromosome XI. *Nature* **369**, 371–378.

Dujon B., Belfort M., Butow R.A., Jacq C., Lemieux C., Perlamma P.S. and Vost M. (1989) Mobile introns: definition of terms and recommended nomenclature. *Gene* **82**, 115–118.

Dunlapp D.D. and Bustamante C. (1989) Images of single-stranded nucleic acids by scanning tunnelling microscopy. *Nature* **342**, 204–206.

Dunn J.J. and Studier F.W. (1983) Complete nucleotide sequence of bacteriophage T7 DNA and the locations of T7 genetic elements. *Journal of Molecular Biology* **166**, 477–535.

Edwards A., Voss H., Rice P., Civitello A., Stegemann J., Schwager C., Zimmerman J., Erfle H., Caskey C.T. and Ansorge W. (1990) Automated DNA sequencing of the human HPRT locus. *Genome* **6**, 593–608.

Edwards J.H. (1994) Comparative genome mapping in mammals. *Current Opinion in Genetics and Development* **4**, 861–867.

Edwards M.D., Stuber C.W. and Wendel J.F. (1987) Molecular-marker-facilitated investigations of quantitative-trait loci in maize. I. Numbers, genomic distribution and types of gene action. *Genetics* **116**, 113–125.

Emmert D.B., Stoehr P.J., Stoesser G. and Cameron G.N. (1994) The European Bioinformatics Institute (EBI) databases. *Nucleic Acids Research* **22**, 3445–3449.

Endo I., Soeda E., Murakami Y. and Nichi K. (1991) Human genome analysis system. *Nature* **352**, 89–90.

Evans G.A. and Lewis K.A. (1989) Physical mapping of complex genomes by cosmid multiplex analysis. *Proceedings of the National Academy of Sciences, USA* **86**, 5030–5034.

Fabret C., Quentin Y., Guiseppi A., Busuttil J., Haiech J. and Denizot F. (1995) Analysis of errors in finished DNA sequences: the surfactin operon of *Bacillus subtilis* as an example. *Microbiology* **141**, 345–350.

Fan J-B., Chikashige Y., Smith C.L., Niwa O., Yanagida M. and Cantor C.R. (1989) Construction of a *Not*I restriction map of the fission yeast *Schizosaccharomyces pombe* genome. *Nucleic Acids Research* **17**, 2801–2818.

Feldmann H. and 96 others (1994) Complete DNA sequence of yeast chromosome II. *EMBO Journal* **13**, 5795–5809.

Ferrin L.J. and Camerini-Otero R.D. (1991) Selective cleavage of human DNA: RecA-assisted restriction endonuclease (RARE) cleavage. *Science* **254**, 1494–1497.

Fields C., Adams M.D., White O. and Venter J.C. (1994) How many genes in the human genome. *Nature Genetics* **7**, 345–346.

Flam F. (1994) Hints of a language in junk DNA. *Science* **266**, 1320.

Foote S., Vollrath D., Hilton A. and Page D.C. (1992) The human Y chromosome: overlapping DNA clones spanning the euchromatic region. *Science* **258**, 60–66.

Franco B. and 15 others (1991) A gene deleted in Kallman's syndrome shares homology with neural cell adhesion and axonal path-finding molecules. *Nature* **353**, 529–536.

Fujii J., Otsu K., Zorgato F., De Leon S., Khanna V.K., Weiler J.E., O'Brien P.J. and MacLennan D.H. (1991) Identification of a mutation in porcine ryanodine receptor associated with malignant hyperthermia. *Science* **253**, 448–451.

Goss S.J. and Harris H. (1975) New method for mapping genes in human chromosomes. *Nature* **255**, 680–684.

References

Goss S.J. and Harris H. (1977) Gene transfer by means of cell fusion II. The mapping of 8 loci on human chromosome 1 by statistical analysis of gene assortment in somatic cell hybrids. *Journal of Cell Science* **25**, 39–57.

Grandbastien M-A. (1992) Retroelements in higher plants. *Trends in Genetics* **8**, 103–108.

Green E.C. and Olson M.V. (1990) Chromosomal region of the cystic fibrosis gene in yeast artificial chromosomes: a model for human genome mapping. *Science* **250**, 94–98.

Gyapay G., Morissette J., Vignal A., Dib C., Fizames C., Millasseau P., Marc S., Bernardi G., Lathrop M. and Weissenbach J. (1994) The 1993–94 Généthon human genetic linkage map. *Nature Genetics* **7**, 246–249.

Haley C.S., Knott S.A. and Elsen J-M. (1994) Mapping quantitative trait loci in crosses between outbred lines using least squares. *Genetics* **136**, 1195–1207.

Hanahan D. and Meselson M. (1980) Plasmid screening at high colony density. *Gene* **10**, 63–67.

Hartl D.L., Ajioka J.W., Cai H., Lohe A.R., Lovoskaya E.R., Smoller D.A. and Duncan I.W. (1992) Towards a *Drosophila* genome map. *Trends in Genetics* **8**, 70–75.

Hartl D.L., Nurminsky D.I., Jones R.W. and Lozovskaya E.R. (1994) Genome structure and evolution in *Drosophila* — applications of the framework P1 map. *Proceedings of the National Academy of Sciences* **91**, 6824–6829.

Hashimoto L. and 13 others (1994) Genetic mapping of a susceptibility locus for insulin-dependent diabetes mellitus on chromosome 11q. *Nature* **371**, 161–164.

Hastbacka J. and 14 others (1994) The diastrophic dysplasia gene encodes a novel sulfate transporter: positional cloning by fine–structure linkage disequilibrium mapping. *Cell* **78**, 1073–1087.

Hauge B.M. and 10 others (1993) An integrated genetic/RFLP map of the *Arabidopsis thaliana* genome. *The Plant Journal* **3**, 745–754.

Hauge B.M., Hanley S., Giraudat J. and Goodman H.M. (1991) Mapping the *Arabidopsis* genome. In *Molecular Biology of Plant Development* (eds. G Jenkins and W. Schurch). Cambridge University Press, Cambridge.

Heale S.M., Stateva L.I. and Oliver S.G. (1994) Introduction of YACs into intact yeast cells by a procedure which shows low levels of recombinagenicity and co-transformation. *Nucleic Acids Research* **22**, 5011–5015.

Hiratsuka J. and 15 others (1989) The complete sequence of the rice (*Oryza sativa*) chloroplast genome: inter-molecular recombination between distinct tRNA genes accounts for a major plastid DNA inversion during the evolution of the cereals. *Molecular and General Genetics* **217**, 185–194.

Hodgson J. (1994) Genome mapping the 'easy way'. *Biotechnology* **12**, 581–584.

Hoelzel R. (1990) The trouble with PCR machines. *Trends in Genetics* **6**, 237–238.

Hoffman E.P., Brown R.H. and Kunkel L.M. (1987) Dystrophin: the protein product of the Duchenne muscular dystrophy locus. *Cell* **51**, 919–928.

Hoheisel J.D. (1994) Application of hybridization techniques to genome mapping and sequencing. *Trends in Genetics* **10**, 79–83.

Hoheisel J.D., Maier E., Mott E., McCarthy L., Grigoriev A.V., Schalkwyk

L.C., Nizetic D., Francis F. and Lehrach H. (1993) High resolution cosmid and P1 maps spanning the 14 Mb genome of the fission yeast *S. pombe*. *Cell* **73**, 109–120.

Huang X.C., Quesada M.A. and Mathies R.A. (1992) DNA sequencing using capillary array electrophoresis. *Analytical Chemistry* **64**, 2149–2154.

Ioannou P.A., Amemiya C.T., Garnes J., Kroisel P.M., Shizuya H., Chen C., Batzer M.A. and de Jong P.J. (1994) A new bacteriophage P1–derived vector for the propagation of large human DNA fragments. *Nature Genetics* **6**, 84–89.

Iris F.J.M. and 10 others (1993) Dense *Alu* clustering and a potential new member of the *NFκB* family within a 90 kilobase HLA class III segment. *Nature Genetics* **3**, 137–144.

Jacob H.J. and 19 others (1995) A genetic linkage map of the laboratory rat, *Rattus norvegicus*. *Nature Genetics* **9**, 63–69.

James M.R. and 16 others (1994) A radiation hybrid map of 506 STS markers spanning human chromosome II. *Nature Genetics* **8**, 70–76.

Jeffreys A.J. and Flavell R.A. (1977) The rabbit beta-globin gene contains a large insert in the coding sequence. *Cell* **12**, 1097–1108.

Johnston M. and 34 others (1994) Complete nucleotide sequence of *Saccharomyces cerevisiae* chromosome VIII. *Science* **265**, 2077–2082.

Jordan B. (1993). *Travelling Around the Human Genome*. John Libbey Eurotext. 188 pp.

Khan A.S., Wilcox A.S., Polymeropoulos M.H., Hopkins J.A., Stevens T.J., Robinson M., Orpana A.K. and Sikela J.M. (1992) Single pass sequencing and physical and genetic mapping of human brain cDNAs. *Nature Genetics* **2**, 180–185.

Kinashi H., Shimaji M. and Sakai A. (1987) Giant linear plasmids in *Streptomyces* which code for antibiotic biosynthesis genes. *Nature* **328**, 454–456.

Kipling D. (1995) *The Telomere*. Oxford University Press, Oxford.

Knoppers B.M. and Chadwick R. (1994) The human genome project: under an international ethical microscope. *Science* **265**, 2035–2036.

Ko M.S.H. (1990) An 'equalized cDNA library' by the reassociation of short double-stranded cDNAs. *Nucleic Acids Research* **18**, 5705–5711.

Kohara Y., Akiyama K. and Isono K. (1987) The physical map of the whole *E. coli* chromosome: application of a new strategy for rapid analysis and sorting of a large genomic library. *Cell* **50**, 495–508.

Konieczny A. and Ausubel F.A. (1993) A procedure for mapping *Arabidopsis* mutations using co-dominant ecotype-specific PCR-based markers. *The Plant Journal* **4**, 403–410.

Koob M. and Szbalski W. (1990) Cleaving yeast and *Escherichia coli* genomes at a single site. *Science* **1250**, 271–273.

Koonin E.V., Bork P. and Sander C. (1994) Yeast chromosome III: new gene functions. *EMBO Journal* **13**, 493–503.

Koop B.F. and Hood L. (1994) Striking sequence similarity over almost 100 kilobases of human and mouse T-cell receptor DNA. *Nature Genetics* **7**, 48–53.

Kostichka A.J., Marchbanks M.L., Brunley R.L., Drossman H. and Smith L.M. (1992) High speed automated DNA sequencing in ultrathin slab gels. *Bio/technology* **10**, 78–81.

Kouprina N., Eldarov M., Moyzis R., Resnick M. and Larionov V. (1994)

References

A model system to assess the integrity of mammalian YACs during transformation and propagation in yeast. *Genomics* **21**, 7–17.

Krishnan B.R., Jamry I., Berg D.E., Berg C.M. and Chaplin D.D. (1995) Construction of a genomic DNA 'feature map' by sequencing from nested deletions: application to the HLA class 1 region. *Nucleic Acids Research* **23**, 117–122.

Kroger M., Wahl R. and Rice P. (1993) Compilation of DNA sequences of *Escherichia coli* (update 1993). *Nucleic Acids Research* **21**, 2973–3000.

Kurata N. and 27 others (1994) A 300 kilobase interval genetic map of rice including 883 expressed sequences. *Nature Genetics* **8**, 365–372.

Lamond A.I. (1988) RNA editing and the mysterious undercover genes of trypanosomatid mitochondria. *Trends in Biochemical Sciences* **13**, 283–284.

Lander E.S. and Botstein D. (1989) Mapping Mendelian factors underlying quantitative traits using RFLP linkage maps. *Genetics* **121**, 185–199.

Lander E.S. and Schork N.J. (1994) Genetic dissection of complex traits. *Science* **265**, 2037–2048.

Lander E.S. and Waterman M.S. (1988) Genomic mapping by fingerprinting random clones: a mathematical analysis. *Genomics* **4**, 231–239.

Larin Z., Monaco A.P. and Lehrach H. (1991) Yeast artificial chromosome libraries containing large inserts from mouse and human DNA. *Proceedings of the National Academy of Sciences, USA* **88**, 4123–4127.

Le Bourgeois P., Lautier M., Mara M. and Ritzenthaler P. (1992) New tools for the physical and genetic mapping of *Lactococcus* species. *Gene* **78**, 29–36.

Lee J–Y., Koi M., Stanbridge E.J., Oshimura M., Kumamoto A.T. and Feinberg A.P. (1994) Simple purification of human chromosomes to homogeneity using Muntjac hybrid cells. *Nature Genetics* **7**, 29–33.

Legouis R. and 14 others (1991) The candidate gene for the X-linked Kallman syndrome encodes a protein related to adhesion molecules. *Cell* **67**, 423–435.

Lehrach H., Drmanac R., Hoheisel J., Larin Z., Lennon G., Monaco A.P., Nizatic D., Zehetner G. and Poustka A. (1990) Hybridization fingerprinting in genome mapping and sequencing. In *Genome Analysis: Genetic and Physical Mapping* Volume 1, (eds K.E. Davies and S.M. Tishman), pp. 39–81. Cold Spring Harbor Laboratory Press.

Lewin B. (1994) *Genes V*. Cell Press, Cambridge, MA/Oxford University Press, New York.

Lichter P., Chang C.J., Call K., Hermanson G., Evans G.A., Housman D. and Ward D. (1990) High-resolution mapping of human chromosome II by *in situ* hybridization with cosmid clones. *Science* **247**, 64–69.

Limbach P.A. Crain P.F. and McCloskey J.A. (1995) Characterization of oligonucleotides and nucleic acids by mass spectrometry. *Current Opinion in Biotechnology* **6**, 96–102.

Ling L.L., Ma N.S-F., Smith D.R., Miller D.D. and Moir D.T. (1993) Reduced occurrence of chimeric YACs in recombination-deficient hosts. *Nucleic Acids Research* **21**, 6045–6046.

Lipshutz R.J. and Fodor S.P.A. (1994) Advanced DNA sequencing technologies. *Current Opinion in Structural Biology* **4**, 376–380.

Lipshutz R.J., Taverner F., Hennessy K., Hartzell G. and Davis R. (1994) DNA sequence confidence estimation. *Genomics* **19**, 417–424.

Lisitsyn N., Lisitsyn N. and Wigler M. (1993) Cloning the difference between two complex genomes. *Science* **259**, 946–951.

Lisitsyn N.A., Segre J.A., Kusumi K., Lisitsyn N.M., Nadeau J.H., Frankel W.N., Wigler M.H. and Lander E.S. (1994) Direct isolation of polymorphic markers linked to a trait by genetically directed representational difference analysis. *Nature Genetics* **6**, 57–63.

Liu S-L., Hessel A. and Sanderson K.E. (1993) Genomic mapping with I-*Cen*I an intron-encoded endonuclease specific for genes for ribosomal RNA, in *Salmonella* spp., *Escherichia coli* and other bacteria. *Proceedings of the National Academy of Sciences, USA* **90**, 6874–6878.

Livshitz M.A., Ivanov I.B., Mirzabelkov A.D. and Florent E.V. (1992) DNA sequencing by hybridization with oligonucleotide matrix (SHOM). Theory of washing out the DNA after hybridization. *Molecular Biology* **6**, 856–865.

Lovett M. (1994) Fishing for complements: finding genes by direct selection. *Trends in Genetics* **10**, 352–357.

Lovett M., Kere J. and Hinton L.M. (1991) Direct selection: a method for the isolation of cDNAs encoded by large genomic regions. *Proceedings of the National Academy of Sciences, USA* **88**, 9628–9633.

Luckey J.A. and Smith L.M. (1993) Optimization of electric field strength for DNA sequencing in capillary gel electrophoresis. *Analytical Chemistry* **65**, 2841–2850.

Ludecke H.J., Senger G., Claussen U. and Horsthemke B. (1989) Cloning defined regions of the human genome by microdissection of banded chromosomes and enzymatic amplification. *Nature* **338**, 348–350.

Lysov Y.P., Khorlin A.A., Khrapko K.R., Shick V.V., Florentiev V.L. and Mirzabekov A.D. (1988) DNA sequencing by hybridization with oligonucleotides: a novel method. *Proceedings of the USSR Academy of Sciences, USA* **303**, 1508–1511.

Maier E., Hoheisel J.D., McCarthy L., Mott R., Grigoriev A.V., Monaco A.P., Larin Z. and Lehrach H. (1992) Complete coverage of the *Schizosaccharomyces pombe* genome in yeast artificial chromosomes. *Nature Genetics* **1**, 273–277.

Marahrens Y. and Stillman B. (1992) A yeast chromosomal origin of DNA replication defined by multiple functional elements. *Science* **255**, 817–823.

Marmur J., Rownd R. and Schildkraut C.L. (1963) Denaturation and renaturation of deoxyribonucleic acid. *Progress in Nucleic Acids Research* **1**, 231–300.

Martin G.B., de Vincente M.C. & Tanksley S.D. (1993) High-resolution linkage analysis and physical characterization of the *Pto* bacterial resistance locus in tomato. *Molecular Plant–Microbe Interactions* **6**, 26–34.

Martin-Gallardo A. and 16 others (1992) Automated DNA sequencing and analysis of 106 kilobases from human chromosome 19q 13.3. *Nature Genetics* **1**, 34–39.

Mashal R.D., Koontz J. and Sklar J. (1995) Detection of mutations by cleavage of DNA heteroduplexes with bacteriophage resolvases. *Nature Genetics* **9**, 177–183.

Maxam A. and Gilbert W. (1977) A new method for sequencing DNA. *Proceedings of the National Academy of Sciences, USA* **74**, 560–564.

Mazur B.J. and Tingey S.V. (1995) Genetic mapping and introgression of genes of agronomic importance. *Current Opinion in Biotechnology* **6**, 175–182.

McClelland M., Jones R., Patel Y. and Nelson M. (1987) Restriction endonucleases for pulsed field mapping of bacterial genomes. *Nucleic Acids Research* **15**, 5985–6005.

McCombie W.R., Adams M.D., Kelley J.M., FitzGerald M.G., Utterback T.R., Khan M., Dubnick M., Kerlavage A.R., Venter J.C. and Fields C. (1992a) *Caenorhabditis elegans* expressed sequence tags identify gene families and potential disease gene homologues. *Nature Genetics* **1**, 124–131.

McCombie W.R. and 20 others (1992b) Expressed genes, *Alu* repeats and polymorphisms in cosmids sequenced from chromosome 4p 16.3. *Nature Genetics* **1**, 348–353.

McDonald J.F. (1993) Evolution and consequences of transposable elements. *Current Opinion in Genetics and Development* **3**, 855–864.

Médigue C., Bouché J.P., Hénaut A. and Danchin A. (1990) Mapping of sequenced genes (1700 kbp) in the restriction map of the *Escherichia coli* chromosome. *Molecular Microbiology* **4**, 169–187.

Médigue C., Viari A., Hénaut A. and Danchin A. (1993) Colibri: a functional database for the *Escherichia coli* genome. *Microbiological Reviews* **57**, 623–654.

Meier–Ewert S., Maier E., Ahmodi A., Curtis J. and Lehrach H. (1993) An automated approach to generating expressed sequence catalogues. *Nature* **361**, 375–376.

Meng X., Benson K., Chada K., Huff E.J. and Schwartz D.C. (1995) Optical mapping of lambda bacteriophage clones using restriction endonucleases. *Nature Genetics* **9**, 432–438.

Mizukami T. and 11 others (1993) A 13 kb resolution cosmid map of the 14 Mb fission yeast genome by non random sequence-tagged site mapping. *Cell* **73**, 121–132.

Monaco A.F. (1994) Isolation of genes from cloned DNA. *Current Opinion in Genetics and Development* **4**, 360–365.

Monaco A.P. and Larin Z. (1994) YACs, BACs, PACs, and MACs: artificial chromosomes as research tools. *Trends in Biotechnology* **12**, 280–286.

Monaco A.P., Neve R.L., Colletti–Feener C., Bertelson C.J., Kurnit D.M. and Kunkel L.M. (1986) Isolation of candidate cDNAs for portions of the Duchenne Muscular Dystrophy gene. *Nature* **323**, 646–650.

Monckton D.G. and Jeffreys A.J. (1993) DNA profiling. *Current Opinion in Biotechnology* **4**, 660–664.

Mott R., Grigoriev A., Maier E., Hoheisel J. and Lehrach H. (1993) Algorithms and software tools for ordering clone libraries: application to the mapping of the genome of *Schizosaccharomyces pombe*. *Nucleic Acids Research* **21**, 1965–1974.

Moyzis R.K., Torney E.C., Meyne J., Buckingham J.M., Wu J-R., Burks C., Sirotkin K.M. and Goad W.B. (1989) The distribution of interspersed repetitive DNA sequences in the human genome. *Genomics* **4**, 273–289.

Murray A.W. and Szostak J.W. (1983) Construction of artificial chromosomes in yeast. *Nature* **305**, 189–193.

Murray J.C. and 26 others (1994) A comprehensive human linkage map with centimorgan density. *Science* **265**, 2049–2054.

Nam H.G., Giraudat J., Dan Boer B., Moonan F., Loos W.D.B., Hauge B.M. and Goodman H.M. (1989) Restriction fragment length polymorphism linkage map of *Arabidopsis thaliana*. *Plant Cell* **1**, 699–705.

Nelson S.F., McCusker J.H., Sander M.A., Kee Y., Modrich P. and Brown P.O. (1993) Genomic mismatch scanning: a new approach to genetic linkage mapping. *Nature Genetics* **4**, 11–17.

O'Brien S.J., Womack J.E., Lyons L.A. Moore K.J., Jenkins N.A. and Copeland N.G. (1993). Anchored reference loci for comparative genome mapping in mammals. *Nature Genetics* **3**, 103–112.

Ohyama K. and 12 others (1986) Chloroplast gene organization deduced from complete sequence of liverwort *Marchantia polymorpha* chloroplast DNA. *Nature* **322**, 572–574.

Okubo K., Hori N., Matoba R., Niiyana T., Fukushima A., Kojima Y. and Matsubara K. (1992) Large scale cDNA sequencing for analysis of quantitative and qualitative aspects of gene expression. *Nature Genetics* **2**, 173–179.

Old R.W. and Primrose S.B. (1994) *Principles of Gene Manipulation*, 5th edn. Blackwell Scientific Publications, Oxford. 474 pp.

Oliver S.G. and 119 others (1992) The complete DNA sequence of yeast chromosome III. *Nature* **357**, 38–46.

Olson M., Hood L., Cantor C. and Botstein D. (1989) A common language for physical mapping of the human genome. *Science* **245**, 1434–1435.

Olson M.V., Dutchik J.E., Graham M.Y., Brodeur G.M., Helms C., Frank M., MacCollin M., Scheinman R. and Frank T. (1986) A random-clone strategy for restriction mapping in yeast. *Proceedings of the National Academy of Sciences, USA* **83**, 7826–7830.

Parimoo S., Patanjali S.R., Shukle H., Chaplin D.D. and Weisman S.M. (1991) cDNA selection: efficient PCR approach for the selection of cDNAs encoded in large chromosomal DNA fragments. *Proceedings of the National Academy of Sciences, USA* **88**, 9623–9627.

Parra I. and Windle B. (1993) High resolution visual mapping of stretched DNA by fluorescent hybridization. *Nature Genetics* **5**, 17–21.

Paterson A.H., Lander E.S., Hewitt J.D., Peterson S., Lincoln S.E. and Tanksley S.D. (1988) Resolution of quantitative traits into Mendelian factors by using a complete linkage map of restriction fragment length polymorphisms. *Nature* **335**, 721–726.

Pease A.C., Solas D., Sullivan E.J., Cronin M.T., Holmes C.P. and Fodor S.P.A. (1994) Light-generated oligonucleotide arrays for rapid DNA sequence analysis. *Proceedings of the National Academy of Science, USA* **91**, 5022–5026.

Pierce J.C., Sauer B. and Sternberg N. (1992) A positive selection vector for cloning high molecular weight DNA by the bacteriophage P1 system: improved cloning efficiency. *Proceedings of the National Academy of Sciences, USA* **89**, 2056–2060.

Polymeropoulos, M.H., Xiao H., Sikela J.M., Adams M., Venter J.C. and Merril C.R. (1993) Chromosomal distribution of 320 genes from a brain cDNA library. *Nature Genetics* **4**, 381–385.

Poustka A., Rohl T.M., Barlow D.P., Frischauf A.M. and Lehrach H. (1987) Construction and use of human chromosome jumping libraries from *Not*I-digested DNA. *Nature* **325**, 353–355.

Prescott D.M. (1994) The DNA of ciliated protozoa. *Microbiological Reviews* **58**, 233–267.

References

Reed P.W. and 13 others (1994) Chromosome-specific microsatellite sets for fluorescence-based, semi-automated genome mapping. *Nature Genetics* 7, 390–395.

Reiter R.S., Williams J.G.K., Feldmann K.A., Rafalsta A., Tingey S.V. and Scolnick P.A. (1992) Global and local genome mapping in *Arabidopsis thaliana* by using recombinant inbred lines and random amplified polymorphic DNAs. *Proceedings of the National Academy of Sciences, USA* 89, 1477–1481.

Rice C.M., Fuchs R., Higgins D.G., Stoehr P.J. and Cameron G.N. (1993) The EMBL data library. *Nucleic Acids Research* 21, 2967–2971.

Rice P.M., Elliston K. and Gribskov M. (1991) DNA. In *Sequence Analysis Primer* (eds M. Gribskov and J. Devereux). Stockton Press, New York.

Richards J.E., Gilliam T.C., Cole J.L., Drumm M.L., Wasmuth J.J., Gusella J.F. and Collins F.S. (1988) Chromosome jumping from DS410(G8) toward the Huntington disease gene. *Proceedings of the National Academy of Sciences, USA* 85, 6437–6441.

Riley M. (1993) Functions of the gene products of *Escherichia coli*. *Microbiological Reviews* 57, 862–952.

Robinson K., Gilbert W. and Church G.M. (1994) Large scale bacterial gene discovery by similarity search. *Nature Genetics* 7, 205–214.

Rommens J.M. and 14 others (1989) Identification of the cystic fibrosis gene: chromosome walking and jumping. *Science* 245, 1059–1065.

Rosenberg M., Przybylska M. and Straus D. (1994) RFLP subtraction: a method for making libraries of polymorphic markers. *Proceedings of the National Academy of Sciences* 91, 6113–6117.

Ross M.T. and 10 others (1992) Selection of a human chromosome 21 enriched YAC sub-library using a chromosome-specific composite probe. *Nature Genetics* 1, 284–290.

Rudd K.E., Miller W., Ostell J. and Benson D.A. (1990) Alignment of *Escherichia coli* K12 DNA sequences to a genomic restriction map. *Nucleic Acids Research* 18, 313–321.

Rudd K.E., Miller W., Werner C., Ostell J., Toltoshev C. and Satterfield S.G. (1991) Mapping sequenced *E. coli* genes by computer: software, strategies and examples. *Nucleic Acids Research* 19, 637–647.

Sanger F., Air G.M., Barrell B.G., Brown N.L., Coulson A.R., Fiddes J.C., Hutchison C.A., Slocombe P.M. and Smith M. (1977a) Nucleotide sequence of bacteriophage ΦX174DNA. *Nature* 265, 687–695.

Sanger F., Nicklen S. and Coulson A.R. (1977b) DNA sequencing with chain terminating inhibitors. *Proceedings of the National Academy of Sciences, USA* 74, 5463–5467.

Sanger F., Coulson A.R., Hong G.F., Hill D.F. and Petersen G.B. (1982) Nucleotide sequence of bacteriophage λ DNA. *Journal of Molecular Biology* 162, 729–773.

Sargent C.A. and 10 others (1993) Cloning of the X-linked glycerol kinase deficiency gene and its identification by sequence comparison to the *Bacillus subtilis* homologue. *Human Molecular Genetics* 2, 97–106.

Schwartz D.C. and Cantor C.R. (1984) Separation of yeast chromosome-sized DNAs by pulsed field gradient gel electrophoresis. *Cell* 37, 67–75.

Schwartz D.C., Li X., Hernandez L.I., Ramnarian S.P., Huff E.J. and Wang Y-K. (1993) Ordered restriction maps of *Saccharomyces cerevisiae* chromosomes constructed by optical mapping. *Science* 202, 110–114.

Schwarzacher T. (1994) Mapping in plants: progress and prospects. *Current Opinion in Genetics and Development* **4**, 868–874.

Sedlacek Z., Konecki D.S., Siebenhaar R., Kioschis P. and Poustka A. (1993) Direct selection of DNA conserved between species. *Nucleic Acids Research* **21**, 3419–3425.

Serikawa T. and 10 others (1992) Rat gene mapping using PCR-analyzed microsatellites. *Genetics* **131**, 701–721.

Shinozaki K. and 22 others (1986) The complete nucleotide sequence of the tobacco chloroplast genome: its gene organization and expression. *EMBO Journal* **5**, 2043–2049.

Shippen D.E. (1993) Telomeres and telomerases. *Current Opinion in Genetics and Development* **3**, 759–763.

Shizuya H., Birren B., Kim U-J., Mancino V., Slepak T., Tachiiri Y. and Simon M. (1992) Cloning and stable maintenance of 300-kilobase-pair fragments of human DNA in *Escherichia coli* using an F-factor-based vector. *Proceedings of the National Academy of Sciences, USA* **89**, 8794–8797.

Sidén-Kiamos I., Saunders R.D.C., Spanos L., Majerus T., Treanear J., Savakas C., Louis C., Glover D.M., Ashburner M. and Kafatos F.C. (1990) Towards a physical map of the *Drosophila melanogaster* genome: mapping of cosmid clones within defined genomic divisions. *Nucleic Acids Research* **18**, 6261–6270.

Sikela J.M. and Auffray C. (1993) Finding new genes faster than ever. *Nature Genetics* **3**, 189–191.

Singer M. and Berg P. (1990) *Genes and Genomes*. Blackwell Scientific Publications, Oxford.

Smith C.L., Econome J.G., Schutt A., Klco S. and Cantor C.R. (1987) A physical map of the *Escherichia coli* K12 genome. *Science* **236**, 1448–1453.

Smith M.W., Holmsen A.L., Wei Y.H., Peterson M. and Evans G.A. (1994) Genomic sequence sampling: a strategy for high resolution sequence-based physical mapping of complex genomes. *Nature Genetics* **7**, 40–47.

Sofia H.J., Burland V., Daniels D.L., Plunkett G. and Blattner F.R. (1994) Analysis of the *Escherichia coli* genome. V. DNA sequence of the region from 76.0 to 81.5 minutes. *Nucleic Acids Research* **22**, 2576–2586.

Southern E.M. (1988) Analyzing polynucleotide sequences. *International Patent Application PCT GB 89/01114*.

Southern E.M., Maskos U. and Elder J.K. (1992) Analyzing and comparing nucleic acid sequences by hybridization to arrays of oligonucleotides: evaluation using experimental models. *Genomics* **13**, 1008–1017.

Sternberg N. (1990) Bacteriophage P1 cloning system for the isolation, amplification and recovery of DNA fragments as large as 100 kilobase pairs. *Proceedings of the National Academy of Sciences, USA* **87**, 103–107.

Sternberg N. (1994) The P1 cloning system — past and future. *Mammalian Genome* **5**, 397–404.

Sulston J., Mallet F., Staden R., Durbin R., Horsnell T. and Coulson A. (1988) Software for genome mapping by fingerprint techniques. *CABIOS* **4**, 125–132.

Sun T-Q., Fernstermacher D.A. and Ves J-M.H. (1994) Human artificial episomal chromosomes for cloning large DNA fragments in human cells.

Nature Genetics **8**, 33–41.

Tanksley S.D., Ganal M.W. and Martin G.B. (1995) Chromosome landing: a paradigm for map-based gene clones in plants with large genomes. *Trends in Genetics* **11**, 63-68.

Thierry A. and Dujon B. (1992) Nested chromosomal fragmentation in yeast using the meganuclease I-*Sce*I: a new method for physical mapping of eukaryotic genomes. *Nucleic Acids Research* **20**, 5625–5631.

Tijssen P. (1993) *Hybridization with Nucleic Acid Probes. Part 1: Theory and Nucleic Acid Preparation.* Elsevier, Amsterdam

Tingey S.V. and Del Tufo J.P. (1993) Genetic analysis with RAPD markers. *Plant Physiology* **101**, 349–352.

Trask B., Christensen M., Fertitta A., Bergmann A., Ashworth L., Branscomb E., Carrano A. and Van den Engh G. (1992) Fluorescence *in situ* hybridization mapping of human chromosome 19: mapping and verification of cosmid contigs formed by random restriction fingerprinting. *Genomics* **14**, 162–167.

Travis G.H. and Sutcliffe J.G. (1988) Phenol emulsion-enhanced DNA-driven subtractive cDNA cloning: isolation of low-abundance monkey cortex-specific mRNAs. *Proceedings of the National Academy of Sciences, USA* **85**, 1696–1700.

Tsai J-Y., Namin-Gonzalez M.L. and Silver L.M. (1994) False association of human ESTs. *Nature Genetics* **8**, 321–322.

Tyler-Smith C. and Willard H.F. (1993) Mammalian chromosome structure. *Current Opinion in Genetics and Development* **3**, 390–397.

Uberbacher E.C. and Mural R.J. (1991) Locating protein-coding regions in human DNA sequences by a multiple sensor-neural network approach. *Proceedings of the National Academy of Sciences, USA* **88**, 11261–11265.

Wada M., Little R.D., Abidi F., Porta G., Labella T., Cooper T., Delle Valle G., D'Urso M. and Schlessinger D. (1990) Human Xq24-Xq28: approaches to mapping with yeast artificial chromosomes. *American Journal of Human Genetics* **46**, 95–105.

Wahl R., Rice P., Rice C.M. and Kroger M. (1994) ECD-a totally integrated database of *Escherichia coli* K12. *Nucleic Acids Research* **22**, 3450–3455.

Walter M.A., Spillett D.J., Thomas P., Weissenbach J. and Goodfellow P.N. (1994) A method for constructing radiation hybrid maps of whole genomes. *Nature Genetics* **7**, 22–28.

Wang G-L., Holsten T.E., Song W-Y., Wang H-P. and Ronald P.C. (1995a) Construction of a rice bacterial artificial chromosome library and identification of clones linked to the Xa-21 disease resistance locus. *The Plant Journal* **7**, 525–533.

Wang Y-K., Huff E.J. and Schwartz D.C. (1995b) Optical mapping of site-directed cleavages on single DNA molecules by the RecA-assisted restriction endonuclease technique. *Proceedings of the National Academy of Sciences* **92**, 165–196.

Wang Y-K., Prade R.A., Griffith J., Timberlake W.E. and Arnold J. (1994) A fast random cost algorithm for physical mapping. *Proceedings of the National Academy of Sciences* **91**, 11094–11098.

Waterston R. and 17 others (1992) A survey of expressed genes in *Caenorhabditis elegans*. *Nature Genetics* **1**, 114–123.

Weatherall D.J. (1991) *The New Genetics and Clinical Practice*, 3rd edn. Oxford University Press, Oxford.

Weinstock K.G., Kirkness E.F., Lee N.H., Earle-Hughes J.A. and Venter J.C. (1994) cDNA sequencing: a means of understanding cellular physiology. *Current Opinion in Biotechnology* 5, 599–603.

Weissenbach J., Gyapay G., Dib C., Vignal A., Morissette J., Millasseau P., Vaysséix G. and Lathrop M. (1992) A second generation linkage map of the human genome based on highly informative microsatellite loci. *Nature* 359, 794–802.

Williams J.G.K., Kubelik A.R., Livak K.J., Rafalski J.A. and Tingey S.V. (1990) DNA polymorphisms amplified by arbitrary primers are useful as genetic markers. *Nucleic Acids Research* 18, 6531–6535.

Williams S.M. and Robbins L.G. (1992) Molecular genetic analysis of *Drosophila* rDNA arrays. *Trends in Genetics* 8, 335–340.

Wilson R. and 54 others (1994) 2.2 Mb of contiguous nucleotide sequence from chromosome III of *C. elegans*. *Nature* 368, 32–38.

Wilson R.K., Koop B.F., Chen C., Halloran N., Sciammis R. and Hood L (1992) Nucleotide sequence analysis of 95 kb near the 3′ end of the murine T-cell receptor α/δ chain locus: strategy and methodology. *Genomics* 13, 1198–1208.

Woo S-S., Jiang J., Gill B.S., Paterson A.H. and Wing R.A. (1994) Construction and characterisation of a bacterial artificial chromosome library of *Sorghum bicolor*. *Nucleic Acids Research* 22, 4922–4931.

Wooster R. and 30 others (1994) Localization of a breast cancer susceptibility gene, BRCA2, to chromosome 13q 12–13. *Science* 265, 2088–2090.

Wu R. and Taylor E. (1971) Nucleotide sequence analysis of DNA. II. Complete nucleotide sequence of the cohesive ends of bacteriophage λ DNA. *Journal Molecular Biology* 57, 491–511.

Youil R., Kemper B.W. and Cotton R.G.H. (1995) Screening for mutations by enzyme mismatch cleavage with T4 endonuclease VII. *Proceedings of the National Academy of Sciences* 92, 87–91.

Zehetner G. and Lehrach H. (1994) The reference library system — sharing biological material and experimental data. *Nature* 367, 489–491.

Zhang P., Schon E.A., Fischer S.G., Cayanis E., Weiss J., Kistler S. and Bourne P.E. (1994) An algorithm based on graph theory for the assembly of contigs in physical mapping of DNA. *CABIOS* 10, 309–317.

Zhang Y., Proenca R., Maffei M., Barone M., Leopold L. and Friedman J.M. (1994) Positional cloning of the mouse *obese* gene and its human homologue. *Nature* 372, 425–432.

Zorio D.A.R., Cheng N.N., Blumenthal T. and Spieth J. (1994) Operons as a common form of chromosomal organization in *C. elegans*. *Nature* 372, 270–272.

Index